Reshaping life

Reshaping life

Key issues in genetic engineering

G. J. V. Nossal and Ross L. Coppel

THIRD EDITION

CAMBRIDGE
UNIVERSITY PRESS

PUBLISHED BY THE PRESS SYNDICATE OF THE UNIVERSITY OF CAMBRIDGE
The Pitt Building, Trumpington Street, Cambridge, United Kingdom

CAMBRIDGE UNIVERSITY PRESS
The Edinburgh Building, Cambridge CB2 RU, UK
40 West 20th Street, New York, NY 10011–4211, USA
477 Williamstown Road, Port Melbourne, vic 3166, Australia
Ruiz de Alarcón 13, 28014 Madrid, Spain
Dock House, The Waterfront, Cape Town 8001, South Africa

http://www.cambridge.org

First published by Melbourne University Press 1984

Printed in Australia

Typeset by Melbourne University Press in 10 point Plantin

A catalogue record for this book is available from the British Library

Library of Congress Cataloguing-in-Publication Data is available

ISBN 0 521 81878 8 hardback
ISBN 0 521 52423 7 paperback

For our children

Contents

Figures

Preface to the third edition

Genetic engineering and related areas of biotechnology continue to advance at an astonishing pace. Since the second edition of *Reshaping Life* was published, there has been a great leap forward in the form of the human genome project, which had been barely discussed then but now has reached virtual completion with the release on Monday 26 June 2000 of a 'first draft' of the full DNA sequence of humans. As a result, the large field of functional genomics has been born. Similar automated approaches to sequencing the genome of pathogenic bacteria and parasites have proven very useful. Study of gene function has been greatly aided by techniques in which genes can be 'knocked out' in mice—this is now a routine research tool. The 'DNA industry' has broadened in a commensurate fashion, now comprising at least 3000 companies.

As research intensifies, so society's concerns have increased. The debate about the use of genetically modified organisms as crops is lively. In 1998, there was a referendum in Switzerland which sought to ban all research on genetically modified animals. Though it was defeated by a two to one majority, it dominated the media for weeks. Subsequently, release of modified plants and contamination of seed stocks has been widely publicized. The extraordinary research leading to the cloned sheep Dolly has raised the possibility, feared by many, that humans may one day be cloned. It is not just DNA research which worries people, but reproductive, cellular and genetic research generally. Much of this technology is being rushed into widespread use, with little time to debate or educate the public. This makes it all the more important that the key issues of modern biotechnology are outlined in a simple yet authoritative way. Powerful technologies can be used wisely or unwisely. Moreover, perceptions as to what is ethically acceptable change with time. While it is prudent to question some uses of the new biology, it is also vital to appreciate the vast potential for good. The third edition seeks to make a contribution to a vital societal debate. At the dawn of the new millennium, we attempt to summarize a complex and rapidly moving field in a way which, we hope, the non-expert can follow with interest.

We have been aided by friends and colleagues in the writing of this book and we would like to acknowledge the assistance of Pamela Dewhurst and Beth Harrison during the writing and revising of this book. The artwork has been supplied by Peter Maltezos and we thank him for his sterling efforts. Finally the time to ponder and write is increasingly precious in the harried environment of today's universities. We would like to thank David Lipman and the staff of the National Center for Biotechnology Information in Washington DC for providing the opportunity to think about where biotechnology and the life sciences have been and where they are going.

GJVN
RLC
Melbourne, November 2001

1

The genie is out of the bottle

When historians look back on the twentieth century, they will conclude that its first half was shaped by the physical sciences but its second by biology. The first half of the century brought the revolution in transportation, communications, mass production technology and the beginnings of the computer age. It also ushered in nuclear weapons and an irrevocable change in the nature of warfare. All these changes and many more rested on physics and chemistry. Biological science, too, was stirring over those decades. The development of **vaccines** and antibiotics and advanced crop **genetics** to feed a hungry world represented proud achievements. Yet the public preoccupation with the physical sciences and technologies, and the immense upheavals in the human condition which these brought, meant that biology and medicine could only move to centre stage somewhat later. Moreover, the intricacies of living structures are such that their deepest secrets could only be revealed after the physical sciences had produced the tools— electron microscopes, radioisotopes and chemical analysers— required for penetrating study. Accordingly, it is only over the last quarter century that the fruits of biological science have jostled their way to the front pages.

In the eye of the storm we find **DNA**. This long but essentially rather simple **molecule** is the key to the puzzle of life. It embodies what each biological **species** looks like, how long it lives, what the limits of its potential are. It specifies in the minutest detail what each plant or animal, and indeed each **cell** in each plant or animal, can do. For this reason DNA has been termed the thread of life; the progressive elucidation of its structure and function have rightly been biology's central preoccupation since 1950. If this is so, why

1

the unique excitement about DNA over the last twenty-five years? What is so new, so special?

For a quarter of a century between 1950 and 1975, DNA was thought by most scientists to be like a remote dictator within a fortified stronghold—inviolate, sacrosanct, issuing orders but itself still and unchanging. Then came the genetic engineers, who with a speed that was truly breathtaking, changed the image of DNA. They made it accessible to a whole new generation of investigators, unafraid of its big reputation. DNA contains the **genes**, the fundamental units of inheritance, the blueprints for the cell's work. Genetic engineers split DNA open, cut out individual genes, transplanted them into **bacteria** or other cells, reproduced them a billion times. They created hybrids in the test tube unlike anything that three and a half billion years of **evolution** had accomplished. Within less than a decade, it became clear that **genetic engineering** and related technologies represented the biggest single advance in the life sciences this century. Genetic engineering held the key to a deeper understanding of human diseases, including cancer. It offered glittering prizes to industry. It promised to free agriculture from constraining requirements for fertilizers and pesticides. There seemed no limits to what the genetic engineers would dare. The genie was out of the bottle.

Unlike the atomic age, which was born in secrecy during a world war, the DNA age began amidst the fiercest blaze of publicity. An unprecedented act of global self-censorship by the scientific community, which in 1974 placed a brief moratorium on genetic engineering research until potential hazards could be assessed, was predictably misinterpreted by society. Scaremongers abounded, and much of the public debate created more heat than light. In the event, patience and sanity prevailed and the research was resumed in 1975 under stringent safeguards that, with hindsight, proved unnecessarily elaborate. It is to the enduring credit of both scientists and their critics that the legislative framework permitting continued advance relied more on regulatory guidelines than on proscriptions and sanctions. As a result, genetic engineering research has infiltrated virtually every nook and cranny of biomedicine. The DNA industry was born and, as we shall see, irrevocably changed the way that pharmaceutical companies did their work. Philosophers and jurists have discovered a new cause, and bioethicists constitute a new professional group. Governments have developed various schemes to nurture the **biotechnology** industry, a source of employment and export income. Above all, journalists of both print and electronic media have exploited the new wonder for stories both of 'breakthroughs' and 'cures', and of conjectural hazards and disasters.

Why, then, another book on genetic engineering? What could possibly remain to be said? The need for this book has dawned on us gradually. In the normal course of work as we interact with many segments of society, many of our lay contacts display a lively interest in research. However, we had been so saturated with gaudy press releases on the one hand, and weighty technical tomes on the other, that we had missed a central point about genetic engineering. Those of us within science who have witnessed the birth of this amazing development, those of us who sit on the endless government committees and write the dry technical reports, even those of us who honestly strive to brief young journalists on their first scientific assignment, have all been too close to the problem. We have forgotten how alien the concepts of genes and cells and molecules are to the layperson. We use our own special language, unconsciously slipping in unfamiliar technical terms, and we soon lose even a hard-working listener. Yet it is vitally important that the potential and also the limitations of genetic engineering be made accessible to a wide public, the more so as its central concepts are really simple. Therefore, the aim of this book is to present the essential elements of genetic engineering within a slim volume in a manner requiring no background in biology and for a readership with no technical expertise in the field. The target audience includes decision-makers at many levels: politicians, financiers and industrialists, community leaders, and academics in non-biological fields. We also hope to stimulate a diversity of people from all walks of life interested in informing themselves about a key development which, slowly but surely, will reshape many aspects of their children's lives. Our greatest wish is to stir the interest of the young, for example the school leaver pondering a career, so that the fascination of the field may tempt a few to become its future devotees. It is not possible to talk about science without using some technical terms. We have tried to keep these to a minimum and have provided simple definitions in the glossary. The first uses of words glossed are given in **bold type** to indicate that they occur in the glossary.

Genetic engineering cannot be intelligently approached without some reference to basic biology. In the next chapter the bare bones of biological organization will be described. Following that, we discuss how genetic engineering works. These two chapters get us over the toughest hurdles; thereafter we deal with practical fruits and social implications of this extraordinary turning point in humanity's affairs. Despite our target of a general audience, the concepts involved in Chapters 2 and 3 are of a more technical nature and may be demanding for readers with a non-scientific background. If a quick scan of them seems formidably daunting to you, we suggest you skip over them and go straight to Chapter 4. This explains the concept of the **genome** and discusses the **human genome project,** one of the most audacious and potentially most rewarding scientific adventures ever essayed. Chapter 5 takes us to the now more familiar ground of how new drugs and products might arise from our new knowledge. This chapter explains how **hormones** and other human **proteins** of use in the treatment of serious diseases can be mass-produced by genetic engineering. Chapter 6 describes how purified genes can help to diagnose hereditary and other diseases, and Chapter 7 goes on to speculate about how, in the future, good genes might be able to be substituted for bad ones right within the body of the patient. We also discuss stem cell therapy and the new developments of **cloning** whole organisms with their awesome implications for the future of our species. Chapter 8 examines some disease problems of the tropical developing countries, with special reference to new vaccines made through genetic engineering. Chapter 9 looks at genetic engineering from an industrialist's viewpoint, and discusses genetically modified organisms and their potential uses in areas such as agriculture, chemistry, mining and waste disposal. The last chapters of the book concentrate on the broader areas of the economics of genetic engineering and some societal issues of the new technology. However, many implications of the new technology have been discussed in Chapters 6, 7 and 9 in particular.

Since the last edition of this book, developments have emphasized the power of recombinant DNA technology and its importance to modern biology. We have seen the molecular study of genes assume absolute centre stage. There is no branch of bio-medical research that has not been profoundly altered by the use of these techniques, and problems that were insoluble a decade ago now routinely yield their secrets. Our understanding of key areas of medicine such as the causes of cancer and of **autoimmune** diseases has been greatly enhanced as a consequence. In this revised edition we cover some of the most exciting developments in basic biology. The discovery of **oncogenes**, genes implicated in the disordered growth characteristic of cancer, and the discovery of **growth factors**, chemical messengers that pass between cells and stimulate growth, are two examples.

A further major development in the last five years has been that this technology has stepped out of the laboratory into the outside world. It is beginning to touch every facet of our lives. Patients are treated with new drugs that were mere laboratory curiosities a short time ago. These proteins, normally present in such small amounts in the body that they could never have been obtained in any other way, are being synthesized and used in routine clinical practice to treat heart disease and other disorders. In the legal arena court cases are being decided on the basis of evidence provided by **genetic fingerprinting**. In commerce, gene companies such as Amgen and Biogen now market products made by recombinant DNA techniques and they are the darlings of the stock exchange. Many food products and additives are produced in increased yield using techniques of biotechnology. New organisms, containing mixtures of genes that as far as we know have never existed before, are in widespread use in agriculture producing substantial amounts of soy bean and corn. We have included sections on these new developments in the book in Chapter 9.

The utilization of these techniques is not without controversy, however. **Transgenic** animals (animals with new genes injected into their cells) have been constructed and we are developing ways to introduce genes into plants and fish. We have allowed limited release of novel micro-organisms into the environment. The technology behind these modified life forms and considerations concerning their release into the world are discussed in Chapter 9. We now have the ability to identify every individual on the basis of their own unique DNA sequences. We will soon have the ability to predict who will be likely to develop heart disease at a young age and who may be cancer-prone. The individual's right to privacy will run smack up against the demands of insurance companies, business and government. Such far-reaching issues require new safeguards to protect the rights of individuals while allowing us to reap the benefits of these techniques. The technology of genetic engineering, like all powerful tools, can be used for good or abused and it is society as a whole that will decide how it is to be applied. We believe that an informed public is a crucial prerequisite to ensuring the technology is used appropriately and in a humane manner.

2

The organization of life

Over three and a half billion years ago, when the conditions of temperature, chemical composition and radiant energy were just right, life appeared on earth. Most probably, this happened in the sea, and on a unique occasion. In other words, the wild profusion of living forms from the humblest of bacteria or algae to humans are all descended from the first primitive living cell. As we shall see, this carries profound implications for genetic engineering. In this chapter we shall describe the cell as the basic common element of all living matter, and will show how cells build up tissues, organs and eventually an individual. Looking deeply into the cell, we will discuss how the components that make up the cell are themselves made. This will bring us to the central role of DNA in the life process and the reasons why different cells in the body perform very different functions.

The cell—a fundamental unit of living matter

One of the utopian dreams of science is to build up a continuum of knowledge about the universe that describes and explains nature in terms of its smallest building blocks. Subatomic particles make up atoms; atoms come together to build molecules; molecules aggregate, often in very complex ways, to build substances, objects, organisms and so forth. Objects and organisms form parts of collective wholes such as families, societies and civilizations. All the matter and life on earth constitutes a planet, which, however, is only part of a solar system in one galaxy of the universe. In practice, the limits of our understanding are such that, while we can make broad statements about how each component or level of complexity relates to the next higher level of organization, many important details remain obscure. In fact, there is a tendency for scientific enquiry to remain within the parameters of a particular level, gaining deeper and deeper insights into it, without worrying too much about connections to the level above or below. So physicists study subatomic particles; chemists, molecules; and geologists, rocks. While interfaces between disciplines sometimes bring forward the most exciting discoveries, most scientists choose not to venture beyond a given frame of reference.

An attempt to introduce biological phenomena to the lay person runs into a similar problem: where to start? What level of organization is the most meaningful entry point to make the excitement of genetic engineering accessible? Most authors step right in and lead with a picture of the molecule, DNA. However, we have chosen to begin with a higher level of complexity, namely the cell. When the cell and its parts have become familiar, the story of the molecules will be much easier to follow. We have tried to pare down to the minimum the amount of technical detail in this chapter. For those who feel uncomfortable with terms such as molecule and protein, the glossary contains a short introduction to such terms.

While there are significant differences between various sorts of cells, there are also sufficient similarities for it to be clear that the cell is the fundamental unit of living matter. In other words, the first life form was a single-celled organism and all subsequently evolved living species consist of a cell or cells. For convenience, Figure 1 represents an animal cell—plant cells or bacterial cells display somewhat different features. A typical animal cell is about one-thousandth of a centimetre (or one two-and-a-half thousandth of an inch) in diameter, which means that a single cell cannot be seen by the naked eye but can be easily recognized under even quite a simple microscope. The limits of the standard microscope are magnification factors of about a thousand, and if one needs to observe cells in still greater detail, a much more complex piece of equipment is required, the electron microscope. This instrument can magnify cells about 100 000-fold and has contributed notably to our present picture of cellular structure and organization. In a complex multicellular organism such as ourselves, it is important to remember that every part—the liver, the brain, the muscles—is made up of cells, conforming to a basic pattern, but differing in details.

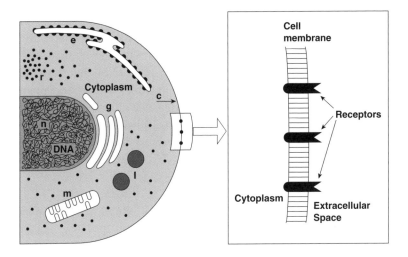

Figure 1. Schematic diagram of a typical animal cell.

The most important organelles are shown: c, cell membrane;
e, endoplasmic reticulum; g, Golgi apparatus; l, lysosomes;
m, mitochondrion; n, nucleus; r, ribosomes.

It is convenient to think of a cell as something like an industrial factory. The first fundamental division of the cell is into two components, the **nucleus** and **cytoplasm**. The nucleus is like the office of the factory. It is the control centre of the cell, the place where the fundamental decisions are made about just what kind of work that particular cell is going to do. The nucleus also contains the apparatus for self-replication of the cell by the process of division that we call **mitosis**. In contrast, the cytoplasm is the shop-floor. It is the place where the work of the cell is done, where the machines and the assembly lines are efficiently positioned and where the power is delivered where it is needed. The cytoplasm is not a random sea of molecules swimming around higgledy-piggledy, but rather a highly organized and compartmented structure. It contains parts which can be seen in the electron microscope and fished out and analysed by sophisticated biophysical and biochemical techniques. These parts are collectively referred to as subcellular

organelles. Only the most important organelles are shown in Figure 1. In the Appendix are brief descriptions of what these organelles do, for the interested reader to follow up. Suffice it to say that within the cell cytoplasm are organelles that provide energy, digestive organelles that break down molecules and most importantly for our story, organelles that make proteins. We shall discuss these protein-making organelles, called **ribosomes,** more fully when we meet **messenger RNA** later in this chapter.

The macromolecules of life

Living cells have plenty of small molecules in them, such as water, salts, sugars, and organic chemicals that perform specialized tasks or act as building blocks. However, somehow it is the large molecules or macromolecules that are most characteristic of living matter. These come in four varieties, namely **nucleic acids,** proteins, carbohydrates and fats. Carbohydrates and fats in the main serve structure and energy-storage functions. Interesting though they are, they need not detain us at this point. The molecules that the budding student of genetic engineering must come to grips with are the nucleic acids and the proteins. Some proteins are also useful for their structural properties, such as the keratin which makes up hair. We are less concerned with these than the so-called globular proteins which perform much more specialized and dynamic tasks in the body.

The proteins

Proteins are really cunningly designed molecular machines which eons of evolution have painstakingly built to be superbly efficient at one task each. Nobody knows exactly how many different sorts of proteins exist in, for example, the human body, but in terms of order of magnitude, a hundred thousand is not a bad guess. Moreover, some proteins come as variations on a theme, the most extreme example being the **antibodies** protecting us against infectious diseases. These really represent only eight sorts of proteins, but tiny, subtle variations in structure make it possible for a person to manufacture a hundred million different antibodies each targeted against a different foreign invader. In the near future, we will know the building block sequence of all the proteins which an individual possesses as information accumulates in the human genome program (Chapter 4). We shall also soon know how these proteins differ in different people, because as we shall see there are may be differences, often quite small ones, in say **haemoglobin** between one person and his or her neighbour. These differences, if they occur, may affect the structure of the protein and how well it can do its job. Thus one form of the protein may function 1% more efficiently in one person, and in a second person, a differing form of that protein may have lost the capacity to do its work at all, resulting in a genetic disease. It is the sum of these differences in proteins which determines differences in functional behaviour of cells, and therefore, eventually, of whole individuals.

The key workhorses of the cell are proteins called **enzymes**. An enzyme is a catalyst, something that encourages a particular chemical reaction to proceed with speed and efficiency. Enzymes are therefore essential for most of the synthetic and degradative processes that go on in the body. Enzymes break down food, help in the transport and storage of energy, are vitally involved in the synthesis of all macromolecules, and obviously are crucial to the replication of the cell. The most important property of an enzyme is its *specificity*. The enzyme which breaks down ordinary table sugar into its component parts does not act on the sugar of milk, which has its own enzyme. Neither one can metabolize the larger storage carbohydrate, glycogen. The specificity of an enzyme depends on its having a particular topological region on its surface with a pattern that is complementary to a corresponding site on the molecule on which the enzyme acts, namely the **substrate**. This congruence of shapes encourages enzyme and substrate to stick to each other when they meet through random molecular motion. Attached to its substrate, the enzyme can do its work. Frequently, the reaction being catalysed actually depends on some different portion of the enzyme molecule than the initial combining site.

In many cases, a particular chemical conversion involves a series or cascade of sequential enzymic reactions. This may mean that a whole row of enzymes is arrayed in the correct spatial order on a specialized membrane framework. Each enzyme does one specific portion of a planned programme of work and then passes on the product to the next enzyme in line, as it does so behaving rather like a machine. So the analogy to an assembly line in a factory is very close.

Other proteins, which are not enzymes, also display the same great specificity. Antibodies, which neutralize poisons and microbes that enter the body, are specific: an anti-measles antibody has no effect against poliomyelitis. Haemoglobin has the specific capacity to take up oxygen in the lungs, where the concentration of oxygen is high, and to release it in the tissues, where the concentration is lower because the oxygen is being used up by the cells. Once again, haemoglobin does this single, vital job by undergoing appropriate changes in shape to snap up and then release the oxygen. So, clearly, the shape of proteins is central to their correct function. A highly sophisticated science called X-ray crystallography has revealed the detailed shape of many proteins down to the exact position of the constituent atoms in relation to one another. The technique can magnify proteins one hundred million times and it is quite fun to walk into an X-ray crystallographer's laboratory and literally walk around a model of a single protein molecule that occupies half a room. Often, the shape gives the clue to how the molecular machine works.

Amino acids: the building blocks for proteins

How does each protein get the special shape it needs to fulfill its mission in life? The answer to that lies in the arrangement of the individual building blocks that make up the protein. Like all the macromolecules, proteins are **polymers** made up of smaller component parts. These small molecules, weighing about one hundred times more than a hydrogen atom, are called **amino acids**. There are twenty different amino acids making up the proteins of living cells. They all have names, of course, but more importantly, each has its own characteristic shape. A typical protein will consist of fifty to two thousand amino acids, with the building blocks arranged in all manner of spatial configurations. You can imagine the vast number of different shapes which that master creator, evolution, has been able to fashion through different arrangements of the twenty different shaped bricks.

One deep truth about proteins must be appreciated before the **genetic code** can be understood. The individual amino acids are coupled together one at a time as a protein is synthesized. A single amino acid attaches to the ribosome to begin the process. Then enzymes couple on the next amino acid, then the next, and so on, one by one, till the full **chain** of amino acids is complete. One can therefore describe a protein as a linear array of amino acids, one after another. This description, however, neglects the fact that the linear array really folds up into a complex, three-dimensional shape. Therefore, as the molecule assumes its final form, amino acid number 3 may find itself just as close, in space, to amino acid 32 or 51 as it is to its neighbours amino acids numbers 2 and 4. Nevertheless, the final shape of the protein under the conditions of the cellular milieu is an obligate consequence of the sequence of amino acids, the *primary structure*, as it is called. The folding follows from the pushes and pulls that the sequential elements of the chain exert on one another. So, in the end, if evolution wishes to tinker with a particular protein, all it has to ensure is a change in order, and a subtle change in final shape will assuredly follow.

The names of the twenty amino acids will not concern us much for the purposes of this book. We shall meet them shortly when considering the genetic code. When protein chemists analyse the sequence of proteins, they frequently make use of one of two conventions when writing up their results. Rather than reporting the structure as a series of full names, they will designate each amino acid by its first three letters, or by a single, conventionally agreed on letter of the alphabet.

The nucleic acids

The other biopolymers or macromolecules of deep concern to us are the nucleic acids. These, too, are composed of building blocks called **nucleotides**. The nucleotides are a little more complicated than the amino acids, because each consists of a **base**, a sugar and a phosphate group. This complexity need not bother us, however, because the sugar–phosphate portion really serves only a structural function. The information content, or coding function, is all tied up with the bases. And now we come to the crucial conceptual difference between the proteins and the nucleic acids. The function of the proteins depends on their final three-dimensional shape. They are machines, working objects. The function of the nucleic acids depends only on the linear order of the bases. They are blueprints or code-books—useful only when translated.

There are two sorts of nucleic acids, DNA and **RNA**, or deoxyribonucleic acid and ribonucleic acid, to dignify them with their full names. The key point about DNA is that it is a string of genes, and each gene contains the coded information for one protein. The DNA is responsible for the master architectural blueprint, held permanently in the head office, the nucleus. RNA comes in various types, but the most important from our point of view, the messenger RNA, is more like a shop drawing—a modified copy of the DNA which actually moves from nucleus to cytoplasm, and is used there, on the shop-floor, where the proteins are made. Both DNA and RNA use only four bases each to perform their coding function. The bases of DNA consist of two molecules with a double organic ring structure, **adenine** and **guanine**, and two with just a single ring structure, **cytosine** and **thymine**. It is a fortunate thing for the sanity of molecular biologists that these four names begin with different letters of the alphabet; they can conveniently be referred to as just A, G, C and T. RNA also uses A, G and C, but instead of T it uses the chemically rather similar base **uracil** or U, which in coding terms is equivalent to T.

The DNA in the nucleus exists as paired strands of circular staircase-like assemblies—the famous double helix of James Watson and Francis Crick (Figure 2). The repeats of sugar and phosphate form the backbones of the strands, and the bases A, G, C and T poke in towards the middle. In fact, the bases of the two strands pair up through a particular type of chemical bond called **hydrogen bonding**. A always pairs up with T and G with C. When it is time for a cell to divide, which obviously requires the whole genetic machinery of the cell to be faithfully copied, the double helix unwinds and enzymes create a new strand complementary to the old one, inserting a T opposite an A, a C opposite a G, an A opposite a T, a G opposite a C and so forth. A simple bacterial cell has just over three million such base pairs in its total DNA, already representing an enormous coding potential. The DNA exists as a single large double-helical molecule. A human cell contains a thousandfold more DNA again! Presumably to facilitate the organization of all this genetic material, the DNA of human cells is broken into forty-six separate chunks called **chromosomes**. Of these, forty-

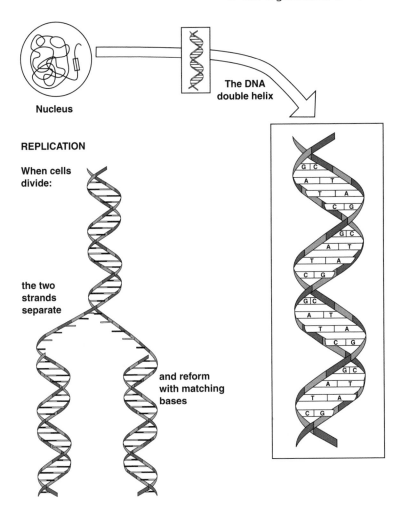

Figure 2. The DNA double helix.

The backbone of each helical strand consists of repeating sugar and phosphate units. The coding function is carried out by four bases: adenine, thymine, guanine and cytosine (A, T, G and C). In double-stranded DNA, the bases are always paired and bonded together in the middle of the helix, A always pairing with T and G with C. When DNA replicates itself before cell division, the two strands of the double helix separate and each strand serves as the template for the construction of a second, complementary strand. Enzymes ensure that, wherever the parent strand to be replicated possesses an A, the incoming building block will be a T, and so forth.

four come as identical pairs, and two are the sex chromosomes, X and X in the female and X and Y in the male. Each chromosome contains a single very long molecule of double-helical DNA, and it also contains some proteins called histones which are tightly bound to the DNA and which may have some regulatory function.

We have said that DNA consists of long chains of four building blocks, the bases, and that the order of the bases along the strands is how genetic information is stored. Part of the DNA is composed of genes that encode the structure of proteins. Interspersed between the genes are stretches of DNA which have no coding function. We still do not really know why all of this non-coding DNA is there, but we do know that the stretches of DNA that abut genes are critically important in controlling how that gene is turned on. In response to signals coming from within or even outside the cell, these control elements (called promotors or enhancers) not only allow messenger RNA synthesis to commence, but also control the number of messenger RNA molecules made. The importance of this is that if we want lots of a particular type of protein, a good way to do this is to make very many messenger RNA molecules, each of which will move out to the cytoplasm, there to direct **protein synthesis**. Similarly, these control elements are crucial to cellular specialization; they will keep the **insulin** gene turned off in a red blood cell but turn it on in the correct cell in the pancreas. When we start to move genes around using the tricks of the genetic engineer, we will swap around these controlling elements as well. In this way we can make a protein in large amounts in a cell that may normally never make any of it at all.

The genetic code

We come now to a concept which confuses many people, but which, essentially, is quite simple. It is that of the genetic code. Consider for a moment the universe of our daily lives. It is filled with objects and entities—buildings, streets, cars, furniture, people, animals. We want to communicate with each other about all these objects and the processes through which they are related. This is the function of language, and moreover we have learnt to record that language in written form. We can regard English as a code that has the following features. The simplest element of the code is a letter, one of twenty-six possible choices in the alphabet. The letters are strung together linearly to form words, most of which are two to fifteen letters long. The words when read by the human eye and brain stand for the real objects and processes that we are trying to describe. In a sense, they gain their reality through the translation process. So it is with the genetic code. Instead of English, with its twenty-six letter alphabet, which, packaged into words, has given us all of Shakespeare and also over two hundred volumes of the *Journal of Immunology*, the genetic code has only four letters in its alphabet, A, G, C and T (or equivalent U). The linear sequence of these four letters in the DNA of each species contains the information for a bee or a sunflower or an elephant or an Albert Einstein. Just as in English, there is a need for punctuation and notations to indicate where words or sentences start and stop. And just as in English, the coded message of the gene gains its reality only when translated in the form of a protein. So the essence of the genetic code is how DNA codes for proteins. Essentially (though there are important exceptions) one gene codes for one protein. This raises the question of how the language of the genes, with its four-letter alphabet, can be translated into the language of proteins with its twenty-letter alphabet. Actually, a cryptographer could come up with a number of elegant solutions to that problem, but nature has found just one, which is universal to each living species. That all living species do use exactly the same code is very important and one of the cornerstones of genetic engineering. This means that if we take a gene from one species, say a human being, and place that gene into a bacterium, the machinery inside that bacterium handles the human gene perfectly. The information will be translated to make a protein whose primary structure is identical to that which would occur in a human cell. Thus genes and their encoded information are completely transportable, from human to animal and from plant to bacteria and potentially of course in the reverse direction.

First of all, the code is linear or sequential. The sequence of bases specifies the sequence of amino acids. Secondly, the code is non-overlapping—that is, a gene segment AGCTGTA can be read either as AGC, TGT, A, or as AG, CTG, TA or as A, GCT, GTA with each base being read once only. Thirdly, it is a triplet code. A sequence of three bases codes for one amino acid, thus AGC codes for an amino acid called serine, while ACG signifies the amino acid threonine, and so forth. Such base triplets are called **codons**, and, given the 4 bases, there are 4 x 4 x 4 or sixty-four codons in all. There are only twenty amino acids to code for, so we have coding information to spare. Three of the codons are used as 'stop' signs, i.e. are punctuation marks. Spare information is obviously available, so most amino acids can, in fact, be coded for by more than one triplet. It turns out that, frequently, the first two letters of the codon are the key ones, and the third does not matter very much. This point is illustrated when we look at a table which spells out the genetic code in total. There is, of course, no need to re-member the details, but the table shows how the genetic code looks.

The four bases of RNA are used—remember U is equivalent to T.

Reading the genetic code

Given these basic elements of the code, how do the events of protein synthesis actually unfold? Signals coming from inside or outside a cell tell the DNA of the genes that it is time for a particular protein to be made. This process of **gene activation** can be likened to turning on an electric light switch. The capacity for all that light was there all along, but only when the switch is turned is the capacity actualized. The signal goes to the control element situated next to the gene and this in turn activates the nearby gene. When a gene is activated (remember: one gene, one protein), the events depicted in Figure 3 unfold. First, an accurate copy of the DNA's genetic message is made in RNA. The DNA acts as a template, and the RNA copy of the gene, still at this moment within the nucleus of the cell, is called a **primary transcript**. The process of copying DNA information into RNA information is referred to as **transcription**. For reasons that are, as yet, by no means clear, the coding information of a gene in cells of higher organisms, including humans, is often interrupted by stretches of bases that have no coding function. These 'nonsense' stretches are referred to as intervening sequences or **introns**, and the portions with coding information are **exons**. The primary RNA transcript is a faithful copy of both exons and introns. Before moving to the cytoplasm, the primary transcript is processed so that the bits corresponding

The genetic code

Codon	Amino acid	Codon	Amino acid	Codon	Amino acid	Codon	Amino acid
UUU	Phenylalanine	CUU	Leucine	AUU	Isoleucine	GUU	Valine
UUC	Phenylalanine	CUC	Leucine	AUC	Isoleucine	GUC	Valine
UUA	Leucine	CUA	Leucine	AUA	Isoleucine	GUA	Valine
UUG	Leucine	CUG	Leucine	AUG	Methionine	GUG	Valine
UCU	Serine	CCU	Proline	ACU	Threonine	GCU	Alanine
UCC	Serine	CCC	Proline	ACC	Threonine	GCC	Alanine
UCA	Serine	CCA	Proline	ACA	Threonine	GCA	Alanine
UCG	Serine	CCG	Proline	ACG	Threonine	GCG	Alanine
UAU	Tyrosine	CAU	Histidine	AAU	Asparagine	GAU	Aspartic acid
UAC	Tyrosine	CAC	Histidine	AAC	Asparagine	GAC	Aspartic acid
UAA	STOP	CAA	Glutamine	AAA	Lysine	GAA	Glutamic acid
UAG	STOP	CAG	Glutamine	AAG	Lysine	GAG	Glutamic acid
UGU	Cysteine	CGU	Arginine	AGU	Serine	GGU	Glycine
UGC	Cysteine	CGC	Arginine	AGC	Serine	GGG	Glycine
UGA	STOP	CGA	Arginine	AGA	Arginine	GGA	Glycine
UGU	Tryptophan	CGG	Arginine	AGG	Arginine	GGG	Glycine

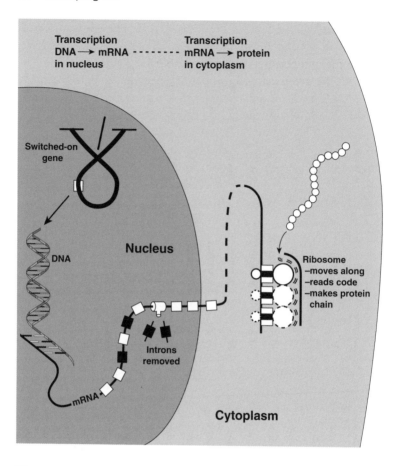

Figure 3. Sequential events in gene activation.

When a gene is activated, the two strands of the DNA double helix separate, and a messenger RNA (mRNA) copy of one of the strands is made. This step is termed transcription. In animal cells, the gene contains stretches of DNA with coding function, called exons, and intervening sections of DNA which do not code for protein and whose function is unknown. These are called introns. As the primary RNA transcript is a faithful copy of DNA, it also contains sequences corresponding to exons and to introns. Inside the nucleus, the messenger RNA is processed so that the intron sequences are cut out and the cuts are sealed up. This processed messenger RNA and a protein chain are progressively assembled, each triplet of three bases coding for one particular amino acid. When the ribosome comes to the end of the messenger RNA, it falls off and the completed protein is released.

to introns are snipped out of the RNA strand, and it is now sealed up to reflect accurately the exons only. The RNA is now ready to move to the cytoplasm as messenger RNA.

Within the cytoplasm, ribosomes attach themselves to the messenger RNA strand and the process of copying the coded information into an amino acid sequence begins. This is called **translation**. You can think of ribosomes as being akin to the head of a magnetic tape machine which reads the messages on the tape as it moves along. In fact, it is a little more complicated that that, because the codons are actually read by *anticodons*, RNA triplets on an adaptor molecule called **transfer RNA**. There are specific transfer RNAs for each amino acid. These transfer RNAs ferry the right amino acid along to the ribosome, and appropriate enzymes link each one to the growing protein chain. As the ribosome moves along to the next codon on the messenger RNA, a new transfer RNA brings up the new amino acid, and so forth. Ribosomes travel along messenger RNA at a set rate. For efficiency, as many as possible will attach themselves, evenly spaced about 100 bases apart, to one RNA molecule, which thereby is read simultaneously by several ribosomes. The longer the message, the more ribosomes crowd on. The size of a **polyribosome** is therefore a reflection of the size of the protein being made.

When the protein chain is complete, it falls off the ribosome and finds itself either free in the cytoplasm or inside a specialized space called the **endoplasmic reticulum**. It folds and coils into the shape already inherent in the amino acid sequence. In some cases, specific enzymes attach sugar molecules to the protein. Furthermore, many important proteins are actually mixtures of several different proteins, and in that case the component proteins are often referred to as *chains*. For example, the insulin molecule consists of one chain of twenty-one amino acids linked to a second chain of thirty amino acids. Antibody molecules consist of four chains, two identical light chains of about 220 amino acids each, and two identical heavy chains which are over twice as long. Clearly, if a protein is to be made by genetic engineering, the need to make more than one chain and hook them together correctly will make the task a little harder.

We have not yet said how the genetic code dictates the structure of other macromolecules, such as carbohydrates or **lipids**. This again is a function served by enzymes. Enzymes, being proteins, are directly coded by DNA genes. Carbohydrates or lipids are built through enzyme action from simple building blocks. If evolution changes even one of the many enzymes involved in their assembly, the final structure will come out somewhat differently. So, albeit indirectly, these macromolecules are under the control of the gene as well.

Cellular specialization as a result of differential gene expression

The chief advantage of multicellular organisms, such as ourselves, is a division of labour, so that different cells can permit themselves the luxury of performing very different and highly specialized tasks. A bacterium, being a single cell, has to have all the enzymic machinery to eat, swim, grow, divide, produce poisons, etc., and it must have that machinery either functioning all the time or ready to go at very short notice. In a human, however, individual cells have to make only a tiny fraction of the proteins that the whole person can make. Stomach cells make digestive juices; thyroid cells make thyroid hormone; pancreas cells make insulin; plasma cells make antibodies; brain cells make the elaborate machinery needed to conduct nerve impulses. Each of these abilities results from the action of proteins, which are in turn available because the gene that carries the instructions is activated in that cell. Clearly, then, only a small proportion of the person's total genes are activated in one particular cell. Presumably, nature could have used one of two strategies to achieve this end. It could have decreed that cells on the way to becoming specialized discarded the genes they did not need. However, this would have allowed little flexibility. A cell could never 'change its mind' about its function. Alternatively, each cell could retain all the genes of the whole organism. The assumption of specialized function would then represent a turning on of certain genes and a turning off of the majority. This second strategy is the one that nature in fact adopted. It means that cellular specialization is the result of differential activation of sets of genes in different cells. The study of gene activation as a key to the process of the formation of specialized organs and tissues is one of the most fascinating and rapidly moving areas of modern cell biology. When it is a little more advanced, it will tell us not only how a liver came to be a liver, a heart a heart and so forth; it will also tell us a great deal more about how they differ, and thus, how they function.

Nature has its own 'plans', its own ways of doing things, but humanity is the wild card in the pack. Before we turn to a consideration of how humanity learnt to trick the genes, let us reiterate what is important for later chapters. Each cell contains the genes of the whole individual, as very long strands of double-helical DNA molecules. The DNA, composed of a linear array of four basic coding units, is copied or transcribed into RNA when a gene is activated. After some processing, this RNA is transported from the nucleus of the cell into its cytoplasm, and there acts as template for the building of proteins. A triplet code ensures that the instructions given in the four-letter alphabet of the DNA and RNA are faithfully translated into the twenty letter alphabet of the proteins. This code is universal and all living creatures use it and pretty well the same translation machinery to make proteins. One gene codes for one protein, and different cells activate different genes, so making different sets of proteins. Therefore each cell is fitted for different tasks.

3

The mechanics of gene transplantation

Now we have not only a reasonable working knowledge of the molecular organization of life but also a feel for DNA, its structure and the powerful ways that the sequence of DNA can be used to answer questions about ourselves. What else can be done with genes and the new genetic technologies, and why would anyone wish to tinker with such perfect machinery? DNA encodes the information for life, RNA copies allow this to be translated into beautiful and useful proteins—who needs more? The answer, as usual in science, revolves around the twin thirsts for knowledge and power. To know more about the genes, we need techniques that can give us a gene in pure form and in sufficient amount for detailed chemical analysis. To harness this knowledge for socially useful purposes, we need methods to transplant the gene to rapidly growing organisms, and to switch it on, as desired, for the mass production of precious proteins. Over the last twenty-five years, engineering techniques of peerless elegance and amazing sophistication have been developed which allow all this and more. Biology and life on this planet will never be the same again.

To give you a feel for how genes themselves can be manipulated to become factories of useful substances, we will discuss some techniques for gene manipulation. Now that the genetic engineering revolution is more than twenty-five years old, there are many different techniques for handling DNA. We have chosen one set of techniques that are relatively easy to understand and give a feel for the problems the genetic engineer must solve in order to harness the power of the gene. As we now know, to get one's hands on a particular gene requires working with DNA. The trouble with the DNA in a cell is that you cannot see the wood for the trees. The molecules are so long and contain thousands of genes. The opening gambit of the genetic engineer is to break the DNA up into small, manageable bits, each containing one or just a few genes. Each little bit, one at a time, is stitched into a special, **virus**-like piece of DNA gifted with the ability for self-replication. These virus-like, recombined DNA molecules now invade rapidly-dividing host cells, again one at a time. Each host cell, frequently a bacterium or a yeast, thereby becomes a factory for one pure gene. Clever tricks allow the genetic engineer to pick out the host cell carrying the gene wanted for that particular experiment. By isolating that one special cell and growing it up to any desired quantity, the one desired gene (or its protein product) can be obtained. So four basic steps are common to all of genetic engineering. First, break the DNA up into short stretches. Secondly, anneal each bit to a suitable ferryman. Thirdly, have these ferrymen invade host cells one at a time, and grow up large numbers. Finally, find the cells with the right transplanted genes in them.

In this chapter we shall uncover the 'nuts and bolts' of genetic engineering technology, describing the key tools and procedures. This will allow us to look at some of the fascinating practical uses in later chapters without pausing too long to worry about methods. If you find the rest of this chapter too technical, skip straight to the summary at the end. Do not think of this as a defeat; even pared down to its bare bones, the engineering is hard to follow in detail, and it would be a pity to miss the other messages of this book through frustration over Chapter 3!

Restriction endonucleases: precision tools par excellence

Consider the human genes: a total of three thousand million base pairs broken up into the chromosomes, each chromosomal double helix still being scores of millions of base pairs long. How to find a given gene? How to study its structure, when there is only a single copy per cell? How to separate it from the other DNA, before transplanting it? The genetic engineer needs precision tools *par excellence* for these endeavours, and the beginning point is a series of enzymes called **restriction endonucleases**.

The restriction endonucleases are enzymes that cut the DNA double helix in very precise ways. They have the capacity to recognize specific base sequences on DNA and then to cut each strand at a given place. This means that the long chain of DNA can be fragmented in a systematic way so that each time it is cut up it yields exactly the same set of fragments. For example, the frequently used enzyme EcoRI recognizes the portion of the double helix reading:

-G–A–A–T–T–C–
-C–T–T–A–A–G–

It makes a very precise cut between the G and the A in each strand, breaking the DNA in two pieces, leaving one end of the DNA looking like this:

-G
-C–T–T–A–A

and the other like this:

A–A–T–T–C–
      G–

Another example is the enzyme HindIII which recognizes the portion of the double helix reading:

-A–A–G–C–T–T–
-T–T–C–G–A–A–

Again a precise cut is made, this time between the A's, giving rise to:

-A
-T–T–C–G–A

on one end and:

A–G–C–T–T–
      A–

on the other.

Obviously, such enzymes will cut a very long DNA double helix many times, in fact every time the particular series of bases turns up in the code. This creates a mixture of fragments of DNA (known as a DNA digest), still in double-helical form, but with a short stretch of single-stranded DNA poking out at both ends of each fragment. These short strands are often termed **sticky ends**. One of the basic features of DNA is that, under suitable chemical conditions, bases 'want to' pair up with each other, A to T, G to C, T to A, C to G. If this is so for a single base pair, it is all the more 'intense a desire' for complementary stretches, so –T–T–A–A very much 'wants to' pair up with A–A–T–T–. Having made the set of cuts with the restriction enzyme, the scientist can set up conditions where the fragments now meet up again, sticky end pairing with its complementary partner. The DNA cuts can be repaired by separate special enzymes called **DNA ligases**. These enzymes act like glue, sticking DNA fragments back into a chain. In the process, of course, it will only occasionally happen that a particular fragment pairs up with the fragment that was its previous neighbour. The likelihood is that, in the mixture of thousands of fragments, a quite random and new arrangement of DNA will result. That is the first basic clue to genetic engineering.

There are now about three hundred restriction endonucleases to choose from, giving the genetic engineer a whole battery of precision instruments. Before we come to what use the genetic engineer makes of these enzymes, we should ask why nature bothered to invent them. In the bacterial world, the enzymes seem to be a protection against foreign DNA that might enter a bacterium and interfere with its function. The bacterium's own DNA is not degraded by its own endonucleases, because it pops a little protective molecule (a methyl group, in fact) on to its own DNA at the recognition site that the restriction enzyme would attack. This blocks cleavage of the bacterium's own DNA, but the restriction enzyme is free to attack other non-methylated DNA that enters, for example the DNA of a virus that attacks the bacterium.

Restriction endonucleases in genetic engineering

The prime use of restriction endonucleases is to break the long DNA double helix into smaller, more manageable bits. Suppose you wished to study an automobile to find out how it functioned. This would be rather difficult if you could not take it to pieces. But by judiciously sorting out the components (gear box, differential, distributor, engine head and so on) and attacking each as an organizational entity, we render the problem more accessible. So it is with DNA. The genetic engineer, beginning to study the structure of a particular gene of interest, does not know where a site for a particular endonuclease might lie. But in trying the various enzymes anyway, the parent DNA molecule is broken into specific fragments that are more readily analysed and manipulated.

Obviously, the longer the stretch of double helix acting as the recognized sequence for a restriction enzyme, the more rarely will it occur, and therefore the bigger the bits of DNA prepared by that particular enzyme. Most of the enzymes recognize stretches of four to six base pairs. If the bases were aligned at random in a stretch a million bases long, a particular sequence of four, say –G–C–G–C–, will occur *on average* once for every $4 \times 4 \times 4 \times 4$ or 256 bases. However, just as in tossing a coin, where on average 'heads' occurs half the time and 'tails' half the time, but strange runs of six heads in a row are not infrequent, it is perfectly possible to find a stretch of fifty bases where –G–C–G–C– turns up twice, or one of a thousand bases where it does not turn up at all. So the pieces of DNA prepared by the relevant enzyme Hha I while *averaging* 256 base pairs in length, are in fact very heterogeneous in length. If the genetic engineer allows the enzyme to act for a long time, until the enzyme has attacked every last recognition site, the million-long molecule of DNA should yield slightly fewer than four thousand different pieces. In practice, it is frequently convenient to allow the restriction endonuclease to act for a sub-optimal length of time, resulting in some –G–C–G–C– stretches *not* being cleaved. Exactly which site is cleaved is largely random so that if you start not with one DNA molecule but with a large number, particular –G–C–G–C– stretches, say forming part of the gene you are interested in, will escape breakage. Thus hetero-geneity of fragmentation is increased by this manoeuvre. A restric-tion endonuclease recognizing six base stretches, such as EcoRI, will find a suitable area for attack only once every 4096 base pairs on average. Again, for incomplete enzymic digestion, the above considerations apply, and the resulting mix of DNA molecules can have some very long fragments. So the genetic engineer can break up DNA in all sorts of ways, and learns by experience which enzymes are best for particular purposes. The principle is simply illustrated in Figure 4.

Having created this heterogeneous mixture of shorter stretches of DNA, the scientist can separate pieces on the basis of their size

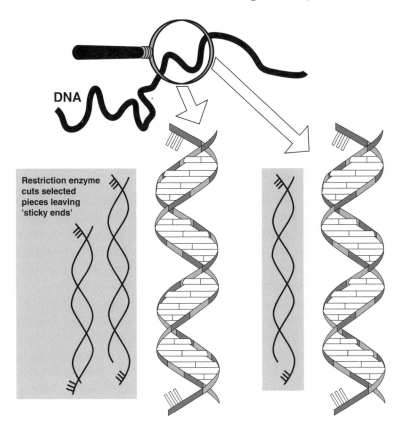

DNA

Restriction enzyme cuts selected pieces leaving 'sticky ends'

Figure 4. The first step in genetic engineering.

DNA to be engineered is broken into small pieces. Restriction endonuclease enzymes recognize a particular sequence and cut the DNA double helix only where such a sequence appears. The resultant shorter stretches of DNA often have 'sticky ends'.

by well-established methods. For example, one can use a kind of molecular sieving technique known as **gel electrophoresis**. In this technique DNA is made to pass through a matrix, somewhat like a jelly, under the influence of an electric current. The bigger DNA pieces keep knocking into the matrix and so get slowed down. The bigger the piece, the more collisions and the slower it goes. So after a period of time the DNA fragments are neatly arranged by size with the smallest bits furthest from the starting point. In this way, scientists can get a kind of fingerprint of a DNA digest. If a fraction of interest is isolated, it can then be attacked by a second restriction enzyme, and if there is a site for that enzyme, a still smaller bit of DNA results. A map of restriction endonuclease sites on a stretch of DNA is a very useful first step in an analysis of its structure.

Vectors for the cloning of DNA in bacteria

We mentioned that the second step after breaking DNA was to stitch each piece into a self-replicating, virus-like stretch of DNA with the capacity to invade certain host cells. As a class, these useful ferrymen for the gene transplanter are known as **vectors** and we must learn a little about them. Vectors are any genetic element that can carry genes about for the genetic engineer.

Bacteria are simple organisms consisting of just one cell. In the popular mind, bacteria are thought of as harmful, disease-causing microbes. In fact, there are millions of species of harmless bacteria occupying particular ecological niches in the biosphere. Many of these are useful, helping to produce vitamins in our intestines, co-operating with plants to allow them to use atmospheric nitrogen as a building block, or fermenting milk into cheese. In fact, human life would be impossible without the bacteria involved in the cycles that move nitrogen, sulphur and oxygen around the biosphere, and allow it to be taken up, used and recycled by all life forms. Bacteria typically have one large circular double-stranded DNA molecule controlling their life and reproduction. As mentioned before, this is about three million base pairs long. Many bacteria also contain smaller double-stranded circular DNA molecules called **plasmids**. These are anywhere from two thousand to a few hundred thousand base pairs long. Plasmids have various curious properties. First, the genes they carry may not be absolutely essential for life and so a plasmid can sometimes leave one bacterial cell and enter another, thereby transferring genetic traits between cells. Secondly, the plasmid can reproduce itself inside the bacterium independently of the main bacterial DNA. Thirdly, a plasmid can sometimes fuse with the main DNA and later work its way loose again, but in such a manner as to drag a piece of the main DNA with it. Nature seems to have evolved plasmids as an efficient way of exchanging genes between bacterial cells. Plasmids are one vital family of vectors for the genetic engineer.

Surprisingly, bacteria also sometimes catch a virus! Viruses are the very smallest forms of life and they are true parasites, being able to live only inside a cell. Bacterial viruses are called **phages** as they sometimes 'eat' the bacterium, at least in the sense of killing it and making it burst. The DNA of a typical phage could be fifty thousand bases long. The virus moves freely from bacterium to bacterium—in fact it possesses a very precise and beautifully designed mechanism for pricking the bacterial wall and injecting its DNA. Phages, also, can sometimes integrate into the main bacterial DNA. When that happens, the virus stops annoying the host bacterium and lies quietly there, replicating only when the bacterium as a whole replicates. Over the years, scientists have worked out tricks to make phages jump into and out of the main DNA chromosome whenever they want. When the phage jumps out, it sometimes carries a host gene or two out with it. Phages are a second family of vectors.

So, these two sorts of self-replicating entities, plasmids and phages, have become key tools for the genetic engineer. Perhaps you have guessed the trick used for the second step in genetic engineering. A population of phages or plasmids is split open by a restriction endonuclease. Pieces of foreign DNA, with appropriate 'sticky ends' prepared through action of the same endonuclease, are added. Frequently, a foreign gene joins up with the plasmid or phage and, with appropriate enzymes, the circle is resealed. Hey presto, you have recombinant DNA! The plasmid or phage is allowed to infect bacteria and so acts as a vector for the foreign gene. Both vector and bacterial host divide as many times as the investigator wishes, and the foreign gene merrily divides with them. Bearing in mind that a bacterium can divide into two every twenty minutes, a billion-fold increase in the foreign DNA can be achieved in ten hours. A variety of vectors exist, tailor-made for genetic engineering, some of them capable of transporting DNA fragments as long as four hundred thousand bases. Placing a vector with a particular, single gene in it into a bacterial cell and then growing billions of progeny is known as **cloning** that gene. A clone is just a technical term for a population descended from a single ancestor by non-sexual means. It is the same word that we use to describe the process of making an identical copy of an animal by non-sexual means. Thus Dolly the sheep is a clone, and scientists are at present examining which other species of animal can be cloned in the same way. We shall return to this issue in a later chapter, but for now we concentrate on the cloning of genes.

Creating a gene library

Let us now see how the cloning of, for example, the genes of a mouse actually takes place. Mouse tissue is minced up and the DNA is chemically purified by techniques that have been known for decades. The DNA is treated by a suitable restriction enzyme, for example, one that potentially produces bits of DNA with an average of 4096 base pairs. If the enzyme acts for a relatively short time, so that it can cut say only one-fifth of the relevant sites, the DNA fragments will, on average, be twenty thousand base pairs long. A suitable vector, say phage virus DNA, is also treated with the same endonuclease, say one producing sticky ends. The two are mixed and then treated with a suitable DNA ligase, a sealing enzyme that 'glues' the DNA strands together. The sticking to- gether and sealing will be an entirely random affair. Sometimes the sticky ends of the vector just join up with each other, re-forming the original, unaltered vector. In other cases, a random bit of the mouse DNA will join up with the two sticky ends of the vector, and the circle reseals with a bit of mouse DNA inserted into the vector to yield recombinant DNA. As the sextuplet of base pairs recognized by the restriction enzyme in our example will not occur regularly and exactly each 4096 base pairs, there will be bits of DNA of various sizes going in, the upper limit being defined by just what each particular vector can accommodate. Figure 5 shows the principle involved.

The scientist is now ready for step three of genetic engineering, namely to insert the vectors into host bacteria one at a time. The genetically engineered vectors are mixed with a population of bacteria that has been grown in a soup-like nutrient broth and then treated to allow the entry of foreign DNA. Some bacteria become infected, but at first the scientist does not know which ones. A useful trick sometimes employed is to ensure that the

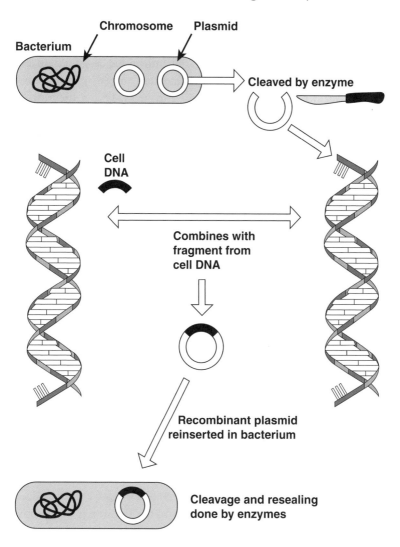

Figure 5. The second step in genetic engineering.

DNA is inserted into a vector. In the case shown, the vector is a bacterial plasmid. The plasmid is cleaved by the same restriction endonuclease as used for the DNA to be transplanted, leaving sticky ends complementary to those of that DNA. Bits of DNA pair up and frequently create a recombined plasmid. DNA cuts are repaired by suitable enzymes.

vector, e.g. a plasmid, carries a gene which confers resistance to a particular antibiotic. Then, in the phase of bacterial growth following the entry of the plasmid, that antibiotic is added to the culture medium, so only bacteria that have taken up the vector (and therefore the resistance gene) can grow. Once infection of bacteria by the vectors has occurred, it is convenient to grow the bacteria not in a soup but on the surface of a jelly. This prevents individual bacteria from swimming around, and allows each bacterium to grow into a **colony** of millions of cells. Colonies are readily visible to the naked eye. Each colony carries a huge number of copies of the vector, and of such foreign DNA as may have been incorporated into it. The general principles are seen in Figure 6. A variant on this general theme is employed with phage vectors. As multiplying phages eventually destroy the bacteria in which they live and infect neighbouring bacteria and destroy them also, one bacterium carrying a phage (in our case, a phage carrying a piece of foreign DNA) can make a hole in a lawn of growing bacteria. These holes are called **plaques**, and can again be spotted with the naked eye.

Let us now do a few simple calculations. In the example we have chosen, the average DNA insert will be twenty thousand pairs long. But the totality of the mouse's DNA, the mouse genome, as it is called, is over three thousand million base pairs. So 150 000 bacterial colonies (or phage plaques) will have within them approximately the total amount of DNA that was in the original mouse genome. Again random chance must be considered. It is quite possible that the 150 000 bacteria received a given stretch of DNA five times over, but that not even one bacterium received another stretch. So, to make reasonably sure that the total genome has somehow found its way into at least one bacterium, genetic engineers will usually prefer to use about ten times more altered bacteria, say 1.5 million in our example, to create the gene collection. Then a given gene will turn up, on average, ten times within the population, and virtually every gene should be present at least once. Moreover, as we have made only a partial DNA digest, even genes which have a restriction site in their middle will be present because we have stopped the enzyme well before it has chewed up all the ten copies of the gene. Such a collection of bacteria or phage plaques is known as a *gene library*. Somewhere within the

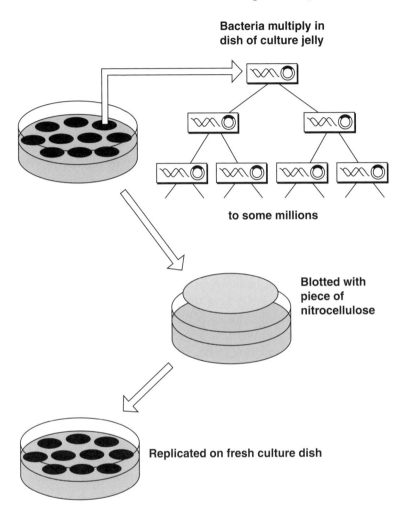

Figure 6. The third step in genetic engineering.

Growth of bacteria carrying recombinant DNA: the plasmids are mixed with host *E. coli* bacteria, and the bacteria carrying recombinant plasmids are grown on the surface of a jellified culture medium. Each bacterium divides until a colony of some millions of bacteria is visible to the naked eye as a spot several millimetres in diameter. By carefully blotting the geographical pattern of colonies on to fresh jellified medium on a second culture dish, using a piece of nitrocellulose, a replica of the original dish can be made. As many replicas as desired can be produced.

library, you will find the gene you want. Just as with an ordinary public library of books, the trick is to find what you want!

What we have just described is a library of *genomic* DNA. There are many occasions when the genetic engineer wants not a genomic library but one reflecting the information content of the messenger RNA of a particular population of cells. Remember, in any one cell type, only a small minority of the genes are active at any one time. Active genes are transcribed into RNA, which is processed in the nucleus and shipped out into the cytoplasm to initiate protein synthesis (Chapter 2). A particular cell's total messenger RNA therefore represents a small subset of the information content of genomic DNA. Moreover, the annoying little stretches of non-coding DNA, the introns, have been cut out during processing of the primary transcript. Suppose you are interested in a particular gene, say the one coding for haemoglobin. It makes good sense to get hold of some cells actively synthesizing haemoglobin, and to purify the messenger RNA from them. If you want to introduce a sophisticated trick, you can even purify the messenger RNA to yield molecules of just the right length to code for a haemoglobin chain. Then that population of RNA molecules can be copied into DNA by an enzyme called **reverse transcriptase**. This copy DNA, or **cDNA** for short, can be cloned into bacterial populations just as described for genomic DNA. A very much smaller library is now sure to have the information you want. You can get even trickier than that. There are ways of getting rid of bits of cDNA that are common to two sorts of cells, and of purifying that of one cell only—a good way of enriching for a specialized cDNA.

Screening the library: how to find the right gene

The genetic engineer is now ready for step four. It is certainly clever to have ways of splitting DNA up into bits, and producing vast amounts of each bit in colonies of bacteria. But how do you find the gene that you want? This is very much a needle in the haystack exercise. In the genomic library, only a few colonies, ten or fewer, amongst our collection of 1.5 million will have the gene we want. As we have seen, the situation is much better for a cDNA library. In devising the screening procedures, the genetic engineers have demonstrated both their genius and their industriousness.

We have already mentioned one clever trick, namely the gene for resistance to an antibiotic which is present in the vector. Another clever trick allows the discrimination of bacteria that have received a *recombinant* vector from ones that have received just an unaltered vector. One particular vector that is commonly used is a plasmid that has in it a gene for a particular protein, β-galactosidase. Now if you feed bacteria with this plasmid a special colourless indicator substance, the β-galactosidase converts it to a coloured dye. The bacterial colony turns blue. Scientists have put a restriction site in the β-galactosidase gene and once cut open, it is here that the foreign DNA is stuck in. If foreign DNA is present it interrupts the sequence of bases that normally code for β-galactosidase with a completely different sequence. Thus translating this gene into a protein produces a nonsense protein that no longer does anything β-galactosidase can. It can no longer make the coloured dye. So now we have a beautifully elegant test for the presence of recombinant DNA molecules. We feed the bacteria the colourless indicator. Bacteria with the unaltered vector turn blue while bacteria with recombinant DNA inside them remain white. Thus, the scientist can concentrate on handling only those bacterial colonies with foreign DNA in them. There are many other ingenious ways to find only bacteria with recombinant DNA in them.

This is the relatively easy part. But what about finding bacteria that have not just any bit of foreign DNA, but the very one you want? We shall consider just two approaches to this, but bear in mind that there are many more. The first involves the 'desire' of complementary DNA strands to bind tightly to each other. Suppose you know the amino acid sequence of the protein for which you are seeking the gene, or even know just one short stretch of sequence data five amino acids long. You look up the genetic code, and synthesize a short stretch of RNA corresponding to that quintuplet of amino acids. Obviously, given the triplet code, this will be fifteen nucleotides long. You arrange the synthesis so that the RNA that is made is highly radioactive. Furthermore, you make sure that, where ambiguities exist for the third nucleotide in the codon, each option is synthesized (see Chapter 2). You now have a radioactive **probe** for the right gene, a labelled RNA molecule capable of binding tightly to the DNA of the gene under appropriate circumstances.

Let us now move to a DNA library which, somewhere within it, has the genetic information you want. Colonies of bacteria containing recombinants are transferred on to a special kind of filter paper, still in their original geographic pattern. The filter is chemically treated to kill the bacteria, break them open, and separate the two strands of all DNA within them. Further treatment removes most of the protein and other materials, but the now single-stranded DNA remains tightly attached to the filter. The radioactive probe is added. Given time, it finds the complementary, well-fitting strand of DNA and binds tightly to it. Obviously the longer the probe, the tighter the fit, and there are ways of getting radioactive ('hot') probes much longer than fifteen in some instances. After the filters are extensively washed to remove unbound amounts of the hot probe, the filter is placed on to an X-ray film in a dark room. The radioactivity of bound hot probe makes a photographic image which shows up as a black spot on the developed film. The investigator then goes back to the original plate of colonies and picks off the one corresponding to the geographical location of the black spot. It should be the one with the right gene in it. Frequently it is necessary to make colonies sit very close to one another to get enough on to a plate, and one might have to regrow the believed right one to make sure it really gave the 'hot spot'. This method is summarized in Figure 7.

The second method is to get the recombinant DNA actually to work for you and synthesize the protein you want. This involved turning on the switch which allows recombinant DNA to be transcribed into RNA and translated into protein as outlined in Chapter 2. All the thousands of genes in the library are activated; each makes large amounts of 'its' protein. Again, replicas are made but this time treated not to release single-stranded DNA, but protein. Then, an antibody to the protein of interest is flooded over the preparation. Antibodies are interesting molecules that the body's natural defense system synthesizes, and they have the capacity to unite specifically with particular proteins. It is fairly simple to make antibodies to insulin, growth hormone, haemoglobin or whatever other precious protein you want. The antibodies identify the colony with the right gene, and the antibodies in turn can be spotted by using a radioactive marker (think of it as a hot anti-antibody) capable of recognizing antibody. The end result again is a dark spot on an X-ray film overlying the area with the 'right' gene.

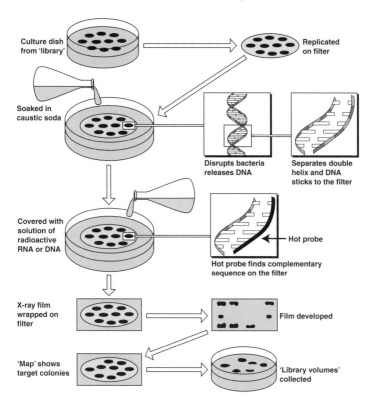

Figure 7. The fourth step in genetic engineering.

An example of finding the gene you want: a library of bacteria containing recombinant plasmids has, somewhere within it, a gene of interest. Replicas of all the culture dishes constituting the library are made on nitrocellulose filters. After colonies of bacteria have grown up, the filters are treated with caustic soda which breaks the bacteria, releasing the DNA, which in turn sticks to the filter. The alkali also separates the two strands of the double helix. After suitable washing, the filters are covered with a solution of radioactive RNA or single-stranded DNA possessing a sequence of a least 14 bases complementary to the gene in question. This 'hot probe' hybridizes to specific sports, corresponding to colonies with the right gene in them. The radioactivity can be detected by placing the filters against an X-ray film in the dark. After several hours or days, the X-ray is developed and a black spot marks the site of the 'right' bacterial colony. The investigator then goes back to the original geographical master pattern and picks off the bacterial colony corresponding to the site of the black spot. This should contain bacteria with plasmids that have taken up the relevant gene.

When screening finds the bacterial clone carrying the right cDNA, the search for the real gene in a total genomic library has taken a giant stride forward. It is easy to make a long stretch of radioactive DNA as a copy of the cDNA, there now being no ambiguities, and this new hot probe can be used as a highly selective screening reagent. Screening the DNA from a million bacterial colonies is tedious, but the technology is improving all the time and the prize well worth the trouble. Furthermore, once a stretch of gene has been identified as being of interest, it is now a relatively simple matter to determine the correct sequence of the chain of bases. If you know the amino acid sequence of the protein, you will soon work out which part of the gene is a coding stretch and which an untranslated stretch or intervening sequence.

Switching on genes: how to get proteins made through genetic engineering

It is one thing to get a gene into bacteria, but another to get those bacteria to act as factories for the manufacture of the gene product, namely the protein for which the gene codes. This is achieved by placing the foreign gene right next to an appropriate control element. For example, bacteria have a complex control system, which years of painstaking research uncovered, that allows them to feed off a sugar called lactose (sugar of milk). As soon as lactose is added to a culture, the gene for the enzyme that breaks it down to the component sugars, galactose and glucose, is switched on. Large amounts of this so-called β-galactosidase enzyme (the same enzyme used for the colour test we described previously) are made by the cell. The area of DNA capable of sensing the presence of the sugar and of turning on galactosidase production is known as the **lac operon**. It is possible to engineer the bacterial lac operon into plasmids, and to insert foreign DNA immediately beside it. When bacteria with such plasmids in them are fed lactose, they make not only the galactosidase enzyme, but also the protein for which the foreign DNA codes. There are a good number of other operons that can be similarly exploited, and much more refined examples than discussed above now exist. Sometimes the 'on' switch can be thrown by a simple temperature change. Vectors that are suitable for switching on genes at will, permitting large-scale synthesis of protein via recombinant DNA technology, are known as **expression vectors**. Some of these are now so sophisticated that bacteria can be forced to make 5 to 10 per cent of their own weight of a specialized protein! As scientists became more skilled at working with different sorts of cells, they were able to make an increasingly diverse set of cells act as protein factories. It is now common to use yeast cells, insect cells and mammalian cell lines from hamsters or monkeys as these factories. In all cases the same general principle, of introducing the gene of interest controlled by an appropriate switch, is followed.

The key tricks of the genetic engineer

The key tricks of the engineer can now be summarized. Use restriction endonucleases to break up long strands of DNA into more manageable bits of one thousand to forty thousand base pairs. Get hold of a plasmid or phage vector capable of entering host bacteria, and open it up by restriction endonuclease treatment. Mix the two and use enzymes to reseal the DNA, creating vectors with foreign, recombinant DNA in them. Allow the vectors to infect a large population of bacteria, and use tricks to reveal which bacteria have successfully become hosts to recombinant DNA. Grow the bacteria up as colonies, or phage populations as visible holes, which can be screened to reveal which one has the DNA sequence of interest to the investigator. Then place this stretch of DNA inside a control system that can be switched on and off at will to create a cheap factory for precious proteins.

For readers more interested in what the genetic engineer can do than in how he or she does it, the important thing to remember is the following. Genes can now be removed from their own home in the nucleus of a cell and transplanted in a myriad ways, either just so that they can be studied in greater detail, or so that they can be activated, or even so that their interaction with other genes can be analysed. This permits a near-infinite variety of experiments and of practical procedures.

4

Genomes: the encyclopaedias of life

In the previous two chapters, we introduced the concept of the gene as the instruction used by the cell to make a protein. We also told you about some of the techniques by which genes can be identified and moved around from organism to organism. When a gene is moved about in this way, the recipient cell is provided with instructions to make a protein that it was unable to make before the gene transfer. As you will recall, proteins are the workhorses of the cell, either acting as enzymes in the synthetic processes that make every component of the cell, or acting as structural components making up the scaffolding of the cell and its various organelles.

Thus, each single gene encodes a single protein, by and large, together with the information that ensures that the protein is made at the appropriate time in the appropriate cell. In addition the gene may contain regions that ensure that the protein is made in response to particular environmental conditions. For example, a simple bacterium growing in solution may be exposed to a sugar such as maltose. The bacterium has an ability to sense its environment and this information is passed to the genes that code for enzymes that metabolize maltose. The genes are switched on and the enzymes to break down this sugar are made, allowing the bacterium to respond appropriately to its environment. Similarly, many of the proteins used by the immune system to combat infection are only made at the time that invading organisms are detected. Special signals will enter the cells of the immune system and signal certain genes that now is the time to make certain proteins that destroy bacteria or viruses.

These examples hint at the complex interplay that occurs in a cell. The bacteria can sense the presence of maltose through the action of specific receptors, proteins on the surface of the bacteria that bind the sugar. The information that maltose is present is signalled to the gene by another set of special signalling proteins. Something similar happens in the cells of the immune system. So, the complex work of a cell is the result of groups of proteins working at the same time, sometimes in synchrony and sometimes as antagonists. There are many, many examples of sets of genes working together, often in response to yet another protein signalling to the regulatory sequences controlling those genes to turn on and make still more proteins.

So exactly how complex can these sets of genes and regulatory sequences get? The answer is that the complexity is very much greater than we can currently understand. Nevertheless, scientists are taking a bold step forward to try and understand the processes of life. They are attempting to obtain the DNA sequence of all the DNA in a number of different organisms. This collection of all the DNA, which includes all the genes of an organism, is called the **genome**. Collectively it encodes all the capabilities of that organism, whether that be the ability to survive in extremes of heat or cold, the ability to digest plastic, or the ability to grow wings and fly. Each species of organism has a unique genome with a unique complement of genes, and it is the differences between total genomes that are responsible for species differences. The genome is passed on from generation to generation and although minor changes occur over time, the bulk of the genome changes very little. Thus when humans have children, we expect those children to be human and not monkeys or mice. They may differ in their temperaments, or heights or hair colour, but largely they will resemble their genetic parents. The differences that they do have are, we believe, the result of two processes. One is the shuffling of genes derived from the parents, to provide a new combination of genes. The second is related to the environment in which that human develops, including the environment inside the mother's uterus.

We will return to the question of how much of human nature is encoded in the genes and how much formed by the environment elsewhere. For now suffice it to say that most of the general properties of human shape, placement of organs, number of fingers and toes and capacity for consciousness result from the correctly co-ordinated action of tens of thousands of proteins encoded by tens of thousands of genes in the human genome. Other organisms will develop different shapes and different characteristics and capabilities because their own genomes have different sets of genes that are regulated in their own precise but unique manner. Surprisingly, the similarities between organisms are often greater than their differences. This is a consequence of the fact that many of the processes required for life are shared between worms, flies and humans. All organisms must have the proteins that allow sugar to be metabolized or DNA and proteins to be synthesized. When nature has come up with a good solution to a problem, that solution is often reused. Thus we can find many genes in the simple, single-celled yeast or in flies or in worms that are very similar to genes present in our own DNA. They apparently encode proteins that perform the same function in all these different cells. Humans and chimpanzees share 98% of their DNA and it is only the other 2% (one-fiftieth of the DNA in the cell) that is responsible for the many differences between the two species. Thus, knowing something about the genes we share and the genes that are unique to us as humans will start to give us an insight into some of the things that uniquely determine the human state.

The human genome project

As Alexander Pope said, 'The proper study of mankind is man', and nowhere has this aphorism been taken more to heart than in the human genome project. Simply stated, the goal of this ambitious program is to determine the nucleotide sequence of all the DNA contained within a human cell. The magnitude of the actual work involved in sequencing all 3.1 billion bases in the human genome is staggering. It has moved biology from a cottage industry into the realm of 'big science'. This is the largest project ever undertaken in the biological sciences and the costs involved are orders of magnitude above those traditionally connected with experiments in biology. The technological demands of the project have pushed the boundaries of current technology in much the same way as the American moon project of the 1960s forced advances in the material sciences and physics. Extraordinary progress has been made over the last few years and milestone has followed milestone. The first billion bases were obtained in November of 1999 and the first complete chromosome, chromosome 22, was completed in December of that year. The 'first draft' of the human genome containing the coding regions of almost all the genes (about 97%) was announced on Monday 26 June 2000. This first draft contains the coding sequence of almost all the genes found in humans. There will be much work to do to fill in the other sequences which will give us information about how the genes are regulated, but even at this early stage, the potential for basic science and for the development of many as yet undreamed of therapeutic agents is enormous.

Let us briefly review the history of this vast international collaboration and look at some of the stages in its evolution. The human genome project (HGP) was conceived in the early to mid-1980s and was widely discussed within the scientific and lay communities through the last half of that decade. There was considerable debate about whether this project was possible or was insanely ambitious. After all, at the time the project was being mooted, all the world's scientists had sequenced a total of five million bases since the beginning of the genetic engineering revolution fifteen or so years previously. The successful completion of the project would require an effort some 1000 times that. It was just not clear that the necessary technology could be developed in a timely manner or that we would even be able to make sense of the sequence once it was obtained. What was clear was that there would need to be advances in the technology of information storage to allow handling of all the data generated, and new software and new generations of supercomputers to fully analyse the information. However the siren song of a great dream proved irresistible and the resources of the global scientific community were marshalled.

In the United States, the project was supported by the Department of Energy initially, and then the National Institutes of Health who were the main research agencies responsible for developing and planning the project. By 1988, the two agencies had agreed to 'coordinate research and technical activities related to the human genome'. In 1990 the two agencies published a joint research plan, *Understanding Our Genetic Inheritance: The U.S. Human Genome Project. The First Five Years FY 1991–1995*. The initiative very rapidly became international with major contributions in the early years from the United Kingdom and France in particular. Other European countries, Australia and some Asian countries also became involved in the project. Major financial support was advanced by governments and charitable foundations such as the Wellcome Trust, and the Human Genome Project is now a truly international effort. Various biotechnology companies also began sequencing the human genome, perhaps most notably the Celera Corporation, a company founded by Dr Craig Venter. He set out to finish the genome as rapidly as possible and in the process discomfited many of the scientists working in the public project. Dr Venter embraced fully the needs for mechanization and robotics and his sequencing project ran as an industrial project rather than a scientific experiment. Making use of public information, but keeping his own results secret, he was able to finish and assemble the human sequence first. However he delayed claiming victory in this race, to allow both public and private projects to simultaneously announce the completion of the first draft sequence. The sequence was finally published in a collection of papers in the important scientific journals *Science* and *Nature* on Monday 12 February 2001.

How does one sequence a genome?

For those of you who are interested, we will spend a bit of time talking about the approaches scientists have taken to sequencing a genome. This section is moderately technical, and readers can safely leave it for another time and move on to the next section without sacrificing much. The message to take away is that the process involves two main phases. The first of these is the determination of the sequence of the DNA by a series of chemical reactions. The second is the storage of the results in a computer and the analysis of the sequence by specially designed computer programs. This second phase is so important that it has spawned its own discipline, the less than euphoniously named 'bioinformatics', which we discuss below.

Our genome sequence is going to be a very long string of the four bases A, C, G and T. What is important is that the order of these is correct. Therefore, the simplest way to sequence a genome would be to take each of the individual chromosomes and start at one end, reading off the bases one by one until we reach the other end. Every base would be in its correct place and the genome would be complete. Such a conceptually simple approach is unfortunately beyond our current state of technology. Instead we must take several large steps backwards in order to reach our goal of a complete genome sequence. This is because currently we can only determine the nucleotide sequence, i.e. the precise order of the bases, of regions of DNA about 650–1000 bases long. Even that is better than the situation some years ago, when the best we could do was 200 or 250 bases in one experiment. So the scientist must attempt to sequence the human genome of 3 billion bases by going at it in these short stretches. The situation is complicated by the fact that it is not feasible to do this systematically and start the next sequence of 650 bases precisely where the previous determination ended.

So the first step is for scientists to divide the genome in sections that are several hundred thousand bases long. These large fragments are cloned into special vectors to make a library of large fragments, much as we described in Chapter 3. These fragments are so long that they are artificial chromosomes in their own right, and in fact they are called YACs (yeast artificial chromosomes) and BACs (bacterial artificial chromosomes). These names reflect the organisms the artificial chromosomes are put into so that many copies are made, and not the source of the DNA cloned into them. Thus we can have BACs containing DNA from humans, mice or flies. Much work is now expended to examine the individual clones and determine how they relate to other clones in the library. A map is developed that lists the clones in the correct order so that collectively they cover the entire chromosome from one end to the other. Genes are mapped to these clones by a process called EST content mapping. The details of this can be found in more technical books on genome sequencing that we list at the web site Reshaping Life (http://www.med.monash.edu.au/reshapinglife). See page 245 for instructions for accessing the site.

Now the next step is to break the DNA into even smaller pieces, by cloning them into plasmid vectors. The process is quite random and the result is thousands and thousands of clones containing short stretches of the genome. We take these clones one by one and we determine the DNA sequence of each small stretch. Because this process is done at random, we might determine the sequence of one part many times and another only once. This process, called 'shotgun sequencing' because of its scattershot nature, is repeated until on average every piece will have been sequenced 5–10 times.

The actual sequence determination is done in a test tube in a chemical reaction in which the DNA one is trying to sequence is copied in a way that makes the newly synthesized strands fluorescent. This process was developed by Professor Fred Sanger of Cambridge University and he received the Nobel Prize for Chemistry in 1980 for this invention. The method has been automated to allow more rapid sequencing but the basic principle is the same as the one he developed. The DNA to be sequenced is first unwound so that it is made into a single-stranded molecule. To this we add a short fragment of DNA called a primer that is complementary to one end of the longer DNA molecule. This primer contains a fluorescent compound which shines brightly if illuminated with laser light. The DNA preparation is then divided into four test tubes which hold the four bases needed to copy the DNA and a special enzyme called DNA-dependent DNA polymerase. This is same copying enzyme that cells use to make a single-stranded DNA into a double-stranded molecule by adding the complementary bases (Chapter 3). Now for the bit of Nobel magic. If this was allowed to proceed on its own, then in each tube we would find many copies of DNA that would now be fluorescent, but would be no wiser as to the sequence. Into each test tube we add a special replication-halting base, but a different one into each tube. In one tube there is a replication-halting base that masquerades for A and can be incorporated into the growing strand at any place that A would normally be (i.e. opposite a T base on the other strand). The difference is that now the DNA chain stops, whereas if a normal base had been there, the DNA chain would have continued to grow. On average it is more likely that a normal base will be incorporated at any particular position, but in a fraction of the reactions, the reaction-halting base A sneaks in and DNA extension stops. Considering all the DNA strands in the tube, some are halted at one place, and some at another, so that in toto, there are examples of the reaction stopping every time there is a T in the opposite strand. In the second tube we place reaction-halting C which terminates the reaction every time G appears in

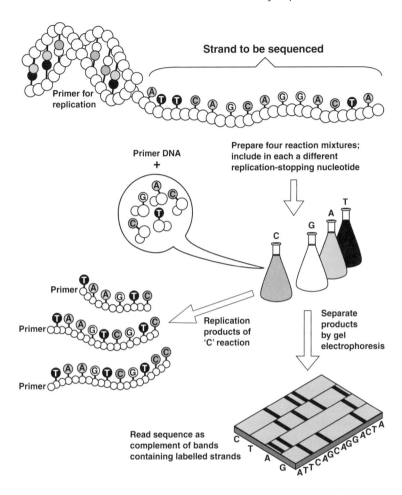

Figure 8. How DNA is sequenced to determine the order of bases.

The DNA strand is first hybridized with a short primer sequence and then an enzymatic reaction is performed in which the primer is elongated using the DNA strand as the template. The reaction mixtures are formulated so that many copies of the strand are made which stop at varying positions during copying. The base at which they stop can be determined and when all the chains are compared side by side, the sequence can be read.

the DNA—and so on for the third tube and reaction-halting G, and the fourth tube and reaction-halting T. We then run these strands of differing length on a gel which sorts them by size and we can see for each position on the lengthening chain, which base caused the reaction to cease. This tells us the nucleotide sequence. The process is illustrated in Figure 8.

These sequences are all fed into a computer running special software and we ask the computer to compare all the sequences and align them so that identical regions overlap. There is a great deal of complexity hidden in that single sentence and it takes a team of dedicated scientists many months to do. The computer tells us if all our shotgun sequences overlap to make a complete YAC sequence. If not, we must do special directed sequencing experiments. Finally, the various YAC sequences are combined, based on the order revealed by our earlier YAC map. We can all breathe a sigh of relief as this genome assembly is over.

The role of bioinformatics

But wait! Now that the physical process of determining the genome sequence is done, we must try to find the genes. What we have produced so far is an enormous amount of raw data that consists of strings of bases, just A, C, G and T, millions or tens of millions or hundreds of millions of characters long. Concealed within this apparently featureless mix are all the genes and regulatory sequences we spoke of earlier. How can we extract meaning from this jumble? The answer is a new discipline in biology mentioned above, that is a mixture of computers, programming and biology. This discipline, called bioinformatics, takes as its remit the development of techniques and programs that will store all of this information now being generated and allow scientists to extract the meaning from it. Organizations such as the National Center for Biotechnology Information in Washington DC and the European Bioinformatics Institute in Cambridge store all of the sequence data generated by genome sequencing and conventional gene sequencing in large databases such as GenBank and EMBL. The sequences are annotated with all the knowledge that scientists now possess about them. Attempts are made to identify the beginning and end of each gene and guesses are made about the function of the genes based on what we have learned about similar genes in other organisms.

The analysis of this information can be quite difficult indeed and supercomputers are routinely employed to try and do as complete a job as possible. To give you a feeling for how complicated this sort of analysis is, imagine taking the Encyclopaedia Britannica and writing out all the individual sentences on little cards. Then take these cards and shuffle them into a random arrangement. Then take each card one by one from the front of the stack and copy all the individual sentences back into one big book. But, before doing so remove all the punctuation, and run all the words together into one long sequence. Just for good measure, while you are doing the copying, throw in any number of extra letters you feel like wherever you wish. Now try and turn this back to your original starting text of the Encyclopaedia.

At first this might seem impossible, but after a little thought some strategies become apparent. You could search this enormous word for longer unusual sequences such as the words Australia or Salmonella or Rousseau and find their location within the text. Your program would have to be clever enough to identify these words interrupted by our random insertions, but in practice this turns out to be quite possible. Then you could examine the surrounding letters and try to match them to known words as well. Examining these in turn, it becomes possible to see the sentence. Once a sentence structure has been discerned, we could lump together all sentences with similar nouns and then try and reform the paragraphs. Clearly much work and trial and error would be involved, but perhaps you can begin to see how the reconstruction process might be achieved. Now analysing genomes is not exactly like this but the process of pulling apart enormous strings of characters and parsing them for some meaning is very similar. We are still in the early stages of applying bioinformatics to the sequence information we have obtained, but we can already see the power of the approach. Much of the most advanced analysis is being done within the walls of the pharmaceutical companies as they examine the genome for potential clues to drugs of great value. This makes trained bioinformaticians highly sought after individuals, particularly those who are skilled both in the computational and the biological aspects of the discipline. This is a career path that aspiring biologists would do well to consider. We have devoted an extensive amount of space to this process of genome sequencing because it is truly a remarkable achievement of the human race and will set the direction of biological science for the next millennium.

What can the genome project teach us?

Identifying the genes within the long mass of bases is an important step forward in the quest to utilize genome data, but it is not the only one. The next challenge for the scientist is to understand what the protein encoded by any particular gene actually does. What a mind-boggling task this is! Consider a jumbo jet with all its individual pieces. Suppose you had the blueprints for such a machine, but instead of these showing the intact aircraft, you had tens of thousands of individual blueprints, each showing a drawing of only a single component, be it bolts or struts or gauges or sections of fuselage. Could you examine all of those single parts, deduce their function and then reconstruct the complete aircraft? A difficult task, and yet reconstructing a human from the blueprints provided by the genome is many orders of magnitude more difficult. How to start?

This general field of determining what genes discovered by computer analysis of genomes actually do, rejoices in the name of 'functional genomics' or often 'post-genomics'. Determining the function of a protein can be a very difficult question to answer indeed, and it could take many years of work by many laboratories to provide answers for just a single protein. A large number of scientists with many different sorts of skills must be conscripted to the fray. Molecular biologists, protein chemists, cell biologists, X-ray crystallographers, physiologists and others all have their parts to play.

Let us consider the case of haemoglobin, the protein that ferries oxygen in our circulation and incidentally the reason why our blood is red. This protein was first purified in crystalline form in 1864 by Ernst Hoppe-Seyler and since then many years of laborious experimentation and many tens of millions of dollars have been expended to try to understand how it works. We now know what haemoglobin looks like, which parts of it carry the oxygen, how it interacts with other proteins and what goes wrong in conditions such as sickle cell anaemia or thalassaemia, where the haemoglobin formed in the body is abnormal. However there are still mysteries associated with this protein, its function and how it is switched on and off. It is daunting indeed to ponder how we could ever hope to have a comparable knowledge of the estimated 50 000–100 000 proteins in the human body. This is a task for the new millennium.

There are many ways to expose the secret of a protein's function, and we shall mention just a few. The first method is an extension of the use of bioinformatics, in which the gene is compared to all the other known genes in all other organisms. If a match is found and the gene you are interested in shares substantial parts of its sequence with some other known gene, this can provide very valuable clues to protein function. Because of this sequence sharing, the resulting proteins probably share some aspects of structure, and as function is based on protein shape, it is also likely that the proteins the two genes encode will have some similarities in function. How similar the proteins are will depend on exactly how much sequence the genes have in common.

Another method to determine protein function that is increasingly used is that of constructing a 'knockout' mouse. Far from being a supremely fit pugilist as the name suggests, these mice are often quite unwell. The reason is that these mice lack a particular gene and therefore cannot make its encoded protein. They develop with this lack and scientists carefully monitor every facet of their development and all their functions to see what has changed as a result of this protein being missing. Sometimes the effects are catastrophic, and sometimes the mice are apparently healthy. In the first instance, the way in which the mice become sick and what systems are affected give insights into the function of the missing protein. Often these mice can be used as models for similar diseases in humans and new therapies can be tested in these mice first. For example, mice that lack a particular protein important in **cystic fibrosis** show very similar signs and symptoms to those unfortunate people with that disease. When knockout mice turn out to be healthy, then it is the scientist who feels unwell at the thought of all that wasted effort and the fact that they are no closer to unravelling the role of their favourite gene. What is often happening in these cases is that a number of proteins may be involved in a particular process and these others can 'cover' for the protein missing in the knockout mouse. Important systems in the body are engineered with a high degree of redundancy because they are so critical for life.

The third method of trying to understand the functions of proteins is called 'DNA microarray analysis' or 'gene chips'. As we have mentioned before, when a gene is turned on to make a protein, it is first copied into a temporary molecule called messenger RNA (mRNA). At any given time within a cell, many different mRNAs are present resulting in many different proteins being made. With the genome sequence in hand, we can construct an array in which every single gene sequence is present. By labelling the mRNA molecules in cells with fluorescent dyes, we can get a visual readout of every gene that is currently making a protein. The way such information is commonly used is to compare cells or organisms in two different states. For example, one could subject cells to some form of treatment, such as a growth factor or a poison, or just the passage of time. Alternatively, one could take cells before and after they have turned cancerous. Comparison of the mRNAs from the before and after states will indicate the genes that are turned on to make protein and often just as importantly, the genes that are turned off. These patterns are complex, but they start to highlight the networks that may be involved in complex processes of life. For example, we have started to learn a great deal about the processes by which cells die. Called apoptosis, cell death is constantly going on in our bodies to maintain cell numbers. It is disturbed in cancer, and we are starting to understand precisely how this occurs.

The next area worth mentioning is that of proteomics. Proteomics is the study of the proteins found in a cell, tissue or organism. At first sight, one might think that this will be little different from studying the genes of an organism, as proteins are encoded by genes. However, there is in fact a further layer of complexity that biologists must be aware of. Once proteins are made, they can be modified further in several ways. They can be cleaved into various pieces so that one gene can make a protein, which is broken down to five or six smaller parts. This often occurs with hormones, with each smaller part having a distinct role in the body. The mode of breakdown may differ in different tissues. Chemicals can be added to proteins, such as lipids or phosphate groups, and these will change the way the proteins function. For example, signals are passed between proteins by the addition of phosphate and the protein with phosphate added (called a phosphorylated protein) is activated and will pass on the message further. Much of what a cell can learn about its external environment is passed from outside the cell to inside the cell by this addition and removal of phosphates to proteins. None of this active signalling process would be obvious from studying genes alone, but by careful examination of the size and electrical charge of a protein, it is possible to see all of the changes we have mentioned here. Finding that a particular protein exists in a cell in two forms differing in size by a phosphate group would lead you to wonder if it might be a signalling protein.

Finally, we should mention structural genomics which is the field that attempts to determine the three-dimensional shape of proteins. As we have mentioned, the function of a protein can often be inferred from aspects of the protein's shape. Further, if we wish to make a drug that interacts with a protein, we are aided by knowing exactly what shape our drug must fit into. There are two ways we have at present of determining protein shape and in both cases we need a source of pure protein at high concentration. Of course genetic engineering can often help us to do this. Then either the protein is solidified and the resulting pure protein crystal exposed to X-rays, or the protein in solution is exposed to radio waves. The way the waves bend around the protein can be analysed and using powerful computers, the shape of the protein inferred. It will be a mighty task to determine the three-dimensional structure of even a small number of proteins in a single genome, but such ambitious projects are being started.

What benefits might flow from knowing the human genome?

It is always dangerous to predict the future, because our predictions largely reveal the poverty of our own imaginations. Let us sketch, however, some of the more obvious benefits of this knowledge, realizing that almost certainly, penetrating insights and great advances will be made in ways we can now only dimly perceive.

There are several areas in which we can expect advances to flow. The production of new drugs and medicines is the most obvious of these and much of this will be covered in Chapter 5. It is worth noting that we will need to know a great deal more than we currently do about how the body functions, before we can truly exploit the information that will become available. At our present state of knowledge, we can most readily use single proteins as drugs. For example, in juvenile diabetes, there is a shortage of insulin in the body. By giving patients insulin, we can improve their condition, although quite terrible problems can still occur. But in other illnesses such as loss of memory, or traumatic amputation of the leg, it is doubtful that a single protein will do much good. Instead, we would need to add a number of proteins, in differing amounts, perhaps combined with other drugs, before we would see restoration of memory or growth of a new limb. We would probably also need to be able to turn off some proteins and would have to administer some form of blocking compounds. At present, we know too little about such complex processes, particularly those that occur as a result of the interplay of many different cell types. Our studies on the genome and the interplay of cellular processes and proteins will provide the basic knowledge by which these more complex therapies will be designed.

The genome provides us with a map that indicates the location of all human genes. This map is extremely useful in finding genes that may be responsible for genetic diseases. Genetic diseases are illnesses that are the result of defects in a single gene. There are hundreds of such illnesses and readers might be familiar with several of the more common ones, such as cystic fibrosis, muscular dystrophy, sickle cell disease and thalassaemia. Often, however, these diseases are quite rare and the total number of affected individuals might be less than one hundred worldwide. Nevertheless, they can give important insights into more common diseases.

Let us illustrate this with an example. Parkinson's disease is a debilitating condition in which spontaneous movement is lost and the sufferers become progressively immobile. It is usually a disease of the elderly, but there is a quite rare condition found in a few families in which the disease appears in much younger people. Researchers have studied DNA from both affected and unaffected members of these families. By carefully examining the DNA using the EST map we spoke about earlier, they were able to identify the region of DNA that was found in patients but present in a different form in unaffected family members. Remember that we have two copies of each chromosome and receive a mixture of DNA from our parents. The Parkinson's sufferers received a piece of DNA with the defective gene while the others received a piece from the other parent containing a normal gene. The region that was mapped as abnormal was a section of chromosome 4. The suspect region contained about 100 genes, but using the available genome data, the researchers picked a gene called alpha synuclein. They found that all the patients had a mistake in the DNA that encoded this protein. In research on other families with early onset Parkinson's disease, the defect was found to be a different one; this time it was a protein involved in the breakdown of alpha synuclein. The same disease was caused in two different ways, but in both cases it led to changes in the level of the same protein, strongly implicating alpha synuclein in the causation of the disease. As this protein also accumulates in brain cells of people with Alzheimer's disease, it may very well be involved in this process as well. We still don't understand its involvement in disease, but researchers now have a way forward to understand not only this rare genetic syndrome but also the more common disease occurring in the elderly. New diagnostic tests and drugs might well result.

In fact the genome sequence can give us insights into many common diseases that are undoubtedly the result of many proteins interacting. Heart disease, high blood pressure, dementia, psychiatric disorders and cancer are complicated conditions and there are many factors, both environmental and genetic, that predispose their appearance. Many people are exposed to the carcinogens in cigarette smoke, but we are all familiar with the healthy octogenarian who attributes his or her longevity to a tot of whisky and a packet of cigarettes a day. Yet we know clearly that the cigarettes that have been consumed are full of cancer-causing chemicals. Why has this person done so well? The answer is complicated. Just as particular diseases run in families, so it appears can good health due to 'good genes', but what exactly are they and how do we find them? The situation becomes more complicated as the number of genes involved in the condition increases so that any individual gene may have quite a weak effect.

Mapping experiments of the type described above for genetic diseases can offer some help, where the gene involved has a very strong effect. Thus the involvement of the genes BRCA1 and BRCA2 in increasing susceptibility to breast cancer was found in this way. However, other methods are required to capture the effect of more minor players. The genome project has faced this challenge by initiating a study that is attempting to identify common differences in DNA sequence among the human population. These variations arise as DNA is copied, and minor changes occur which from generation to generation become fixed in the DNA of different people. The variations can be additions or losses of bases or changes in the bases called **mutations**. As long as these changes don't catastrophically affect the ability of an important protein to do its job, then they are tolerated and passed on to one's children. Indeed most of the mutations are found in DNA outside gene-coding regions. These mutations or **single-nucleotide polymorphisms** (SNPs—pronounced snips—as scientists refer to them) are very common, occurring perhaps one in every 300–500 bases. They can be used as markers to identify different regions of DNA in population studies. Thus, for example, high blood pressure might be found in a population study to be associated with a particular SNP. One or several of the genes that are near this SNP are likely to be involved in causing high blood pressure. Much work will be needed to build up a database of hundreds of thousands of SNPs, but once this has been created the search for multiple genes involved in complex conditions becomes much easier. We will discuss in Chapter 6 the societal ramifications of the ability to predict someone's likely future health on the basis of their inherited genes.

One very positive outcome of a detailed understanding of genetic variation will be an improvement in the capacity to deliver existing treatments. We are all familiar with the fact that currently available medicines seem to work very well for some people but not for others. Some people develop stomach ulcers when taking aspirin, even in quite low doses, whereas others do not. Tablets for high blood pressure make some people dizzy but not others. Some people taking digitalis get symptoms of toxicity such as nausea and vomiting at dosages which in other people are barely enough for the drug to have any action whatsoever. Many of these differences in responsiveness are related to differences in how people metabolize drugs, and to the causes of their disease state. When we understand the genome and genetic variation better, doctors will be able to tailor their treatments and predict which of several alternatives will work best in a particular patient with fewest side effects.

It is obvious that the human genome project will have enormous implications for how medical care is delivered. There are also a number of very legitimate fears about genetic testing and how it may be used. A new area spawned by these worries is called ELSI—Ethical, Legal and Social Impacts—and these concerns form much of the subject matter of this book and of the web site Reshaping Life (http://www.med.monash.edu.au/reshapinglife). The site also includes an updated section on some of the findings about the human genome (see page 245 for instructions on accessing the site).

Other genomes and their lessons

We have concentrated on talking about the human genome, how that vast project will be completed and what the outcomes may be. But there are a number of other genome projects going on in which the DNA of other organisms is being completely sequenced. Genome projects have been commenced to sequence the genome of mice, worms, weeds, important agricultural plants such as rice, and of bacteria and other harmful organisms. Quite a large number of genomes have already been completed. The first of these was *Haemophilus influenzae*, the cause of a number of serious childhood infections. This achievement demonstrated that sequencing of microbial genomes was possible, although the actual strain of *Haemophilus* chosen was a laboratory enfeebled strain, so much of what enables this microbe to cause disease is still to be discovered. The second genome was that of *Mycoplasma genitalium*, one of the simplest of living organisms with a genome size of 580 000 bases containing 468 genes. This represents about the minimal set of genes that allow an organism to live without being parasitic on something else. Although viruses can get away with just a handful of genes, they require a host cell to do all the work of replicating the virus. If you are going to make your way as a free-living organism, it seems you can't do with many fewer than those 400 plus proteins. Compare this with the complexity of humans whose genome has 3 000 000 000 base pairs and who possess over 30 000 genes.

As we start the new millennium, scientists have completed over 60 genomes. In almost all cases, these are genomes of simple single-cell organisms such as bacteria or yeast. Some of these bacteria are pathogens, i.e. they are capable of causing disease in other organisms, including ourselves. Indeed a major focus has been to determine the genome of many of the bacteria that are important causes of disease in people. Truly, scientists have an insatiable desire to understand life in all its majesty and intricacy.

So what use is sequencing these other genomes, apart from being a vast stamp-collecting exercise? Well it turns out that these genome sequences not only give profound insights into how these other organisms function, but can also be used to help us understand ourselves. As alluded to earlier, we can see the close relatives of our genes in the genes of many other species. In these simpler species we can do experiments that will give us a clue to what the gene does, experiments that are not possible in humans. We have already discussed knockout mice and the insights they provide, but it is also possible to do very powerful experiments in worms and fruit flies that tell us most important things about ourselves and the ways our bodies are controlled.

Let us focus for a moment on the pathogen sequencing programme. This programme is based on the premise that if we know all the genes found in an organism and understand the function of the proteins they make, then we know all it can ever do. Thus identifying its weaknesses becomes easier. Here is one example. Sequencing projects have told us that the malaria parasite, a major scourge of humans that kills millions every year, has a particular organelle called a plastid. The DNA that encodes the proteins of the plastid is very close to that found in plants. These proteins in plants can be interfered with by particular chemicals that are used to kill weeds. Scientists are now finding that malaria parasites are also susceptible to these same herbicides and malaria can be cured using herbicides that are quite safe in people. Thus, genomics has given us new drugs from a most unexpected source. It would have taken many years of painstaking trial and error drug testing to have come up with the same result, assuming that anyone would have had the wacky idea of trying to cure malaria with weedkillers.

Pathogen genome sequencing also gives us extraordinary history lessons as to how these dangerous organisms arose. It turns out that bacteria can very easily swap DNA about. They pick up DNA from other bacteria, from viruses that infect them and even from free DNA floating by. Sometimes this DNA is useless and sometimes it encodes proteins that are harmful to the bacteria. Occasionally, though, this DNA enables the bacteria to do new things, such as survive in a different environment, be stickier or metabolize previously undigestible substances. If these new abilities help the bacteria survive then they will be passed on to the next generation and become widespread. We can see from sequencing many bacterial genomes that pathogens frequently pick up a parcel of genes that makes the bacteria more dangerous to humans. Called **pathogenicity islands**, these sets of genes encode antibiotic resistance or the ability to invade human cells. Many bacteria that cause important diseases such as cholera, typhoid fever and food poisoning picked up gene parcels that turned previously harmless bacteria into major killers. Knowing this, however, gives us a set of targets against which we might design new drugs. If we can inactivate these proteins involved in causing disease, then these bacteria can be rendered harmless. We can peacefully co-exist with them as we do the trillions of harmless bacteria that peacefully reside on and in our bodies.

But there are still more treasures to be found from sequencing microscopic life. Bacteria have colonized the world in its entirety and can live in environments that are totally inhospitable for other life forms including humans. Bacteria live under the sea in magma vents where temperatures exceed 100°C, in hot springs and in salt deserts. They can metabolize substances we find poisonous such as metal ions, or organic solvents such as those used in dry cleaning. Some bacterial strains can tolerate doses of radiation thousands of times higher than mammals can. Their genomes will provide us with instruction kits on how to do these difficult things. One could imagine taking the genes for digesting organic pollutants and placing them into radiation-resistant bacteria. These could then be used for decontaminating radioactive waste dumps. Other bacteria could be constructed to render harmless many of the quite toxic metals such as mercury. This activity, which is called bioremediation, is an active focus of research for the United States Department of Energy which has supported sequencing of the genomes of many of these versatile bacteria. The bacterial proteins that function at high temperatures could be purified and used as catalysts in an industrial process that typically runs at temperatures that destroy ordinary proteins. There are many exciting possibilities.

Beyond this is one of fascinating problems of science. What is it that specifies life? Is there a minimal set of activities that a cell must carry out and what are these? How many genes are truly essential for life? We know that many viruses have between three and ten genes but they are not really alive, as they must rely on another cell to replicate. We have mentioned *Mycoplasma genitalium*, with its 468 genes. This is the lowest number we know of in a free-living bacterium, but can organisms survive with fewer genes? One way to approach this issue is to examine the genomes of a large number of bacteria. If a gene is truly essential for life, then it will be present in all the bacteria sequenced. Genes that enable particular bacteria to do a special thing such as metabolize metals or cause whooping cough will be confined to bacteria with that particular talent. Such comparisons are starting to be made. Eugene Koonin at the National Centre for Biotechnology has predicted that a self-replicating bacterium could exist with 256 particular genes. Scientists could, with current technology, stitch together versions of these genes from a number of bacteria and create a life form that has never existed in nature. We think there are very few people who believe that this experiment should be allowed to be performed in our current state of knowledge. Much debate and the passage of time is needed before we should consider the prospect of making life forms *de novo*.

To conclude, the age of genomics is upon us and biology and human civilization will never be the same again. In the succeeding chapters, we will discuss several ways in which this technology can be exploited. As with all powerful technology, there is a potential for misuse and harm. It is essential that the power of this technology be harnessed for our benefit. We will discuss this is greater detail in succeeding chapters.

5

Factories for precious proteins

The lot of the nineteenth-century physician could not have been an easy one. A great deal of knowledge about the natural history of disease had accumulated. Diagnoses, though not resting on today's vast infrastructure of specialized tests, were frequently accurately established through listening to the symptoms and spotting a few clinical manifestations. So, doctors knew the havoc that disease could wreak in people's lives, but could do little about it! True, there were powerful preparations like digitalis, opiates and belladonna, helpful in a limited way, but all too often nature just had to be allowed to take its course.

The revolution in the capacity of scientific medicine to intervene and secure the prevention or cure of disease rests essentially on four developments: vaccines, replacements, operations and drugs. Vaccines, which harness nature's own defence system to prevent specific infections, have combined with more sanitary ways of living and have helped to rid the developed countries of many plagues. Replacement therapy, be it by way of vitamins, blood transfusions or hormones like insulin, has allowed the control of diseases where some bodily organ underperforms or has been destroyed. Surgery, including modern obstetrics, can be proud of its many triumphs. But in thinking of biotechnology we should look closely at drugs; for it is the large armamentarium of these powerful pharmaceuticals which provides today's physicians with the chief weapons of daily medical practice.

Taken in the broadest sense, drugs are small organic molecules that have the capacity either to mimic or to interfere with some important biological process. Some are natural products, like many of the antibiotics. Increasingly, however, drugs represent synthetic products tailor-made to have a particular detailed shape. Organic chemists think up and make molecules with particular characteristics, say to block bodily chemicals that cause acid production in the stomach, or to mimic the natural molecules that encourage muscles in the bronchial passages to relax. If the idea works, you have a new drug for peptic ulcers or for asthma, though years of trial and error involving lengthy testing in experimental animals must precede any human application. Much of the chemical signalling system of the body, between nerve and nerve, nerve and muscle or cell and cell, itself depends on localized release of small organic molecules, so it stands to reason that other small molecules can be introduced to affect the various control loops, and to put them right where they have gone wrong. The twin sciences of pharmacology (the study of drug action) and therapeutics (the use of drugs in treatment of disease) therefore rest on a firm scientific foundation.

Small molecules alone are not enough, however. There is a further area of scientific medicine which is newer, still uncertain, rapidly developing and full of exciting potential. It involves the use of larger molecules in therapy. Many of these are proteins, and the term 'biologicals' has come to connote therapeutically useful substances of this sort. The need for biologicals can be explained simply enough. Powerful though interventions based on small molecules may be, much of the body's day-to-day business is transacted via macromolecules. Frequently things go wrong that can only be remedied by large molecules, with their greater degree of specialization of function. In fact, some biologicals are not particularly new. We have already briefly mentioned insulin, a life-saving discovery made nearly eighty years ago. Then there are various blood products such as albumin, useful to replace lost body fluids, and gamma globulin, which can prevent some infections, though only for a time. These substances are relatively easy to prepare, because they are abundant, in the pancreas and in blood respectively. The problem with many therapeutically useful biologicals is that they are present in blood or tissues in low concentrations, so that extracting enough becomes a difficult and costly exercise. And, of course, this is where genetic engineering comes to the rescue.

Once a gene for a protein has been inserted into bacteria or yeast or a mammalian cell using a suitable expression vector, it matters not one whit whether the protein for which that gene codes is an abundant one, or one of exceeding preciousness because it is so rare in tissue or body fluids. Properly engineered, the gene for the rare protein can be made to work just as hard and just as fast as that for a common protein. This does not just mean a considerable cost saving in preparation of proteins compared with extraction from natural sources. In many cases, it means an all or none difference—something that was not available at all before is now potentially available in large amounts. Even for proteins that were available before, genetic engineering is frequently a much more practical and sensible way to proceed. Several examples will illustrate the point.

Somatostatin cloned: the start of a long adventure

There is a hormone called somatostatin, which perhaps enjoys more fame than it deserves. Somatostatin is a small protein, only fourteen amino acids long. It is made in the pancreas and elsewhere by specialized cells, and it represents one of those control or feedback loops of which nature is so fond. Its chief role is to counterbalance the growth-promoting effects of pituitary growth hormone on storage of foodstuffs in the body. Somatostatin will be remembered as the first protein made by *E. coli* through genetic engineering. It is stunning to recall that this feat was only achieved as recently as 1977. How extraordinary that such mind-stretching feats can appear so commonplace so soon! In those early days of genetic engineering, it seemed easiest not to find the gene for somatostatin, e.g. in a cDNA library, but rather to synthesize it! As there were only fourteen amino acids in the protein, a stretch of forty-two bases would code for the whole lot. The organic chemistry needed to place forty-two nucleotides into the right sequence was far from simple in those days, but it was achieved by the scientists Keichi Itakura, Francisco Bolivar and Herbert Boyer. They were able to place this synthetic gene into a plasmid vector and obtain synthesis of quite respectable amounts of somatostatin. In the process it became apparent that a few special tricks would have to be learnt to make the technique work perfectly. It was necessary to devise clever ways of splitting the desired protein from the galactosidase enzyme that had been used as the switching device inside the bacterium (see Chapter 3). Also, a fair proportion of the transformed bacteria degraded the newly formed, genetically engineered protein almost as fast as they made it, and there seemed to be relatively little rhyme or reason determining when this happened. Indeed, this tendency for intracellular breakdown remains a major nuisance in genetic engineering research to this day, but it is being combated with progressively greater success and predictability. It looks as though partially completed fragments of precious proteins, made when only a part of the gene has been transplanted, are especially susceptible to enzymic degradation. Ways to increase yields include adding inhibitors of the enzymes that digest proteins to the bacterial growth media or using bacteria that lack degradative enzymes. Tremendous strides in what one might term the production technology side of genetic engineering have been made since 1977, and further improvements in the near future are certain.

Human growth hormone: a classical milestone for genetic engineering

David Goeddel and colleagues from the genetic engineering firm Genentech announced in October 1979 their success in the manufacture of human growth hormone. Three factors distinguished this feat from the somatostatin achievement. First, this hormone is fourteen times bigger than somatostatin, being 191 amino acids in length. Secondly, on this occasion the gene was not first synthesized, but rather fished out of a gene library by what has now become one of the classic processes. Messenger RNA was purified and put into a library inside *E. coli* bacteria, the right gene was found within the library and engineered into a suitable expression vector (a plasmid) and large amounts of hormone were made. Thirdly, while the somatostatin work of Itakura had been supported by industry, the growth hormone work was completed within industry, or at least so it was claimed at the time. We will return to this area shortly, when we highlight some of difficulties in the relationship between researchers in academia and those in biotech companies.

What is the importance of human growth hormone? It is one of the most important triggers for cellular and tissue growth. A tumour of the pituitary cells making growth hormone sometimes causes an overproduction of this vital substance. The end result is a giant, with particular overgrowth of hands, feet and chin. Such acromegalic individuals have a characteristic appearance; the former heavyweight boxing champion of the world, Primo Carnera, was a classic example. In contrast, if these hormone-producing cells of the pituitary fail, or are destroyed by disease, the result is what is called a pituitary dwarf. Fortunately, this is a rare condition. Until recently, nothing could be done about it. Then, a process was developed in which human pituitaries were collected from post-mortem rooms, and, by a laborious chemical technique, human growth hormone was extracted from them. Given the small size of the human pituitary, it is no wonder that the hormone prepared in this way was extremely expensive. Regular injections of this laboriously collected growth hormone succeeded in restoring the potential dwarf to normal height. Then tragedy struck. Three patients who had been treated with pituitary extract in the 1950s and 60s developed Creutzfeldt-Jakob disease. This is a rare but fatal disease of the brain caused by a very unusual type of infectious agent, a prion, that takes many years to cause disease. People with this disease become confused, lose their memory and slowly die as their brain loses all ability to function. One of the pituitary donors had died of the disease, not then recognized as being infectious, and the prions were passed on in the extract, lying dormant, only to strike with devastating effect thirty years later. Use of the pituitary extract was banned. Fortunately, the recombinant product was becoming available and it was rushed into production to provide the required treatment. At least three companies now manufacture recombinant growth hormone and the price of treatment has decreased somewhat. In market terms, of course, pituitary dwarfism constitutes a very limited outlet for sales. However a cheap and pure source of human growth hormone will allow experimentation with, and conceivably use of, growth hormone in other conditions where rapid cellular growth is desirable—such as wound healing, and repair of burns and fractures. Growth hormone also has the effect of selectively breaking down fat tissue in preference to muscle so that people on diets who are treated with growth hormone will lose weight only from fatty regions. The use of such a drug as a cosmetic is an area of considerable controversy. The market here is huge and companies will be tempted to explore these more unusual uses of biologicals. This is an area in which one would like to hasten slowly, but the pressures of commercialization are becoming increasingly strong and it is already clear that many of the new biologicals are being used in surprising and sometimes illegal ways. The central significance of this technical feat of expressing growth hormone is that it showed how well genetic engineering can work in a 'real life' setting.

There is an extraordinary aspect of the story of the cloning of human growth hormone that is being examined within the court system. Regardless of how the case develops it highlights some of the pressures and temptations now existing in an industry where large amounts of money can be made or lost. We started this section by mentioning David Goeddel and scientists at Genentech and their announcement in 1979 of the cloning of human growth hormone. This announcement was greeted with great enthusiasm and made Genentech a much wealthier company. Sales of human growth hormone, marketed under the name Protropin, have totalled more than US$2.5 billion since the product's release. The story of the cloning of growth hormone had commenced at the University of California in the laboratory of Howard Goodman, and scientists there including Peter Seeburg had successfully identified clones containing the growth hormone sequence. However these clones were not able to synthesize any growth hormone protein and thus were useless for treating anyone. The University of California however did patent their discovery in 1977, and this patent included claims of ownership of the sequence of the growth hormone gene. Dr Seeburg left the university and joined Genentech where he and David Goeddel attempted to make clones expressing the growth hormone protein. According to testimony given in the trial, these attempts ran into trouble, so one night Dr Seeburg returned to the laboratory at the university, at a time he expected Dr Goodman to be absent, and took some clones that he had generated earlier in his research. Subsequently Genentech successfully made growth hormone, but Dr Seeburg claimed that this success was critically dependent on the clones he had taken from the university. The University of California sued Genentech for infringement of patent and claimed a considerable amount of money in damages. The history of the cloning of growth hormone was examined in minute detail in court and much conflicting testimony was heard as to whether clones were removed or whether they helped Genentech achieve production of growth hormone. In the event, the jury was unable to reach a unanimous decision and was deadlocked with eight of the nine jurors finding in favour of the university. Before a second trial could be held, the two sides settled with Genentech paying US$200 million to the university, but denied any infringement of patent or appropriation of research material. The university inventors and their collaborators John Baxter, Howard Goodman, Joseph Martial, Peter Seeburg and John Shine shared US$85 million. Whatever the truth in this particular case, the story reinforces the fact that there are huge profits to be made in biotechnology and that there are great temptations. The traditional role of universities as dispassionate searchers after knowledge is being changed irrevocably. We will return to this tension between the orderly process of scientific inquiry and the imperatives of a profit-driven industry many times throughout this book.

Human insulin through recombinant DNA technology: a potentially large market

Sugar diabetes, or diabetes mellitus, is really two separate diseases. In one form ('late onset diabetes'), which is common enough to affect up to 2 per cent of the population, there is a relatively mild failure to cope with the load of sugar that is taken into the body with a carbohydrate-rich meal. This failure is not due to a failure of insulin secretion from the pancreas gland. In fact insulin levels are normal or even raised. The disease has to do with how insulin is used by the body. This form of diabetes usually starts in middle to late life, is frequently accompanied by obesity and does not require insulin injections. It can be managed either by restricting carbohydrates in the diet or by pills of active antidiabetic agents, which combat the resistance to insulin and thus cause a lowering of blood sugar levels. The second form of diabetes, which occurs perhaps one-third as often, is much more serious. It often, though by no means always, starts in teenage life. It is due to a destruction of certain cells in the pancreas which are specialized for insulin secretion. Untreated, it is usually rapidly fatal. Since 1923, when Banting and Best discovered that this disease is due to a failure of insulin formation, it has been treatable and millions of lives have been saved. However, the price has usually been two or even up to four insulin injections per day. The commercial source of insulin is from the pancreas of either cattle or pigs. While pharmaceutical companies have become immensely skilled at purifying the insulin present in such glands collected from the abattoirs, the present approach can still lead to difficulties for some patients. Both bovine and porcine insulin, though very similar to human insulin, are just slightly different from it in their composition. The immune defence system of the body is devilishly clever at picking up even minute differences between 'self' and 'not self'. A proportion of diabetic patients make antibody against insulin, because of these small differences between the animal and the human product. This has two deleterious consequences. First, the antibodies may neutralize insulin's action, thus requiring much greater amounts to be injected. Secondly, when insulin and antibody against it meet in the tissues, a nasty inflammatory response is set up, and so the site where the patient injects himself/herself becomes red, swollen and very painful. You can imagine the agony if this happens twice a day, every day! Thirdly and more speculatively, diabetics even if well-treated get a number of serious late complications, and it cannot be excluded that some of these are associated with the injection of foreign protein. So human insulin through recombinant DNA technology could represent a highly significant advance. Also, from the point of view of the pharmaceutical companies, it represents a far more significant sales volume than the other hormones mentioned.

Human insulin through genetic engineering presented scientists with yet another problem. Unlike growth hormone, insulin is really made up of *two* proteins joined together by an organic bridge called a **disulphide bond**. Both of these protein subunits of insulin are made in the pancreas by a complicated process. First the cell makes a molecule over twice as big as the final insulin and then chops bits off it in a highly organized manner that ensures the correct combination of the two subunits. Obviously an engineered bacterium would not possess the highly specialized enzymic machinery to do this bit of processing. Between 1977 and 1980 several groups found different ways of solving this problem, and the example is of immense importance for the future, as many important proteins are in fact composite molecules—multichain structures, as they are termed technically. This fact would no longer be a barrier to their successful manufacture by genetic engineering. In fact, suitably engineered bacteria can now make so much insulin that they literally bulge with it, as can be observed with the electron microscope. This insulin, used in long-term clinical trials, has been shown to be safe and effective and is now available commercially. Not all the expected advantages of this product over animal sources of insulin have materialized. Long-term complications of diabetes still occur and some patients are allergic to the human form of insulin. However, a compelling new reason for preferring the genetically engineered form of insulin has arisen. Bovine spongiform encephalopathy or 'mad cow disease', a transmissible condition due to those unusual agents prions we mentioned earlier, has sensitized the public to the dangers of using products produced from farm animals. It should be stressed that there is no evidence that insulin of pig or cattle origin is contaminated or dangerous to humans. Nevertheless, as a general principle it is safer to administer a genetically engineered product. Thus we would expect to see the gradual disappearance of all alternative forms of insulin. In time recombinant insulin should be much cheaper as the patent for its use expires and other companies market competing products.

Haemophilia and the problems of blood transfusion

Perhaps the most dramatic example of the worth of replacing collected proteins with recombinant proteins is in the area of transfusion of particular proteins found in blood. Haemophilia is a hereditary disease caused by the lack of a factor involved in blood clotting called Factor VIII. This can be purified from fresh human blood by a series of fractionation steps, and has been generally available to treat haemophilia for some time now. As a result, the lives of tens of thousands of people have been improved immeasurably.

Alas, a major problem then arose with the appearance of the disease known as **AIDS**, the acquired immune deficiency syndrome. This disease, with which we are now all too familiar, has affected tens of millions of people around the world, many of whom have died. The disease is due to infection by the human immunodeficiency virus (HIV-1) that attacks some of the white blood cells involved in immunity. People with the disease are susceptible to cancer and to many organisms that are harmless to normal individuals. The virus is spread by sex, or by transfer of contaminated blood, even in small amounts, as can happen through needle-sharing by drug users. Tragically haemophiliacs, who receive transfusions or extracts of blood pooled from thousands of different donors, had a greatly increased risk of contracting the disease. Before adequate screening tests for AIDS, many patients were infected. Even now with universal screening of blood donations (at least in Western countries), there remains a time period of about three weeks in which infected individuals cannot be diagnosed by the standard antibody test and thus their blood could be used by a blood bank. The risk of any individual acquiring AIDS from infected blood donors is minute (estimated at one in a million) when the blood is being properly screened, but for the haemophiliacs who must receive blood components pooled from many people, it is obviously much larger, though clever ways of removing the virus do exist. Recombinant Factor VIII would eliminate any risk of acquiring AIDS and ensure plentiful supplies of this essential medicine for all haemophiliacs. We should also keep in mind that HIV virus is just one of several possible infectious agents that can contaminate the blood supply. Readers may have heard of the various hepatitis viruses including hepatitis B and hepatitis C that can also cause major health problems. The risk of catching these and other as yet unidentified viruses found in blood will be eliminated by the use of recombinant protein. The production of Factor VIII was a formidable task as it is an extremely large protein of complicated structure with many sugar groups attached to the protein. Using mammalian cells, several groups were able to express Factor VIII and this protein is now available for use in the clinic. Its arrival signifies that we have now have the technology to produce essentially any protein we wish, probably regardless of what organism it comes from.

Molecular regulators of growth

So far, we have dealt with proteins, the therapeutic potential of which was fairly well established before the availability of genetically engineered material. However, the promise of genetic engineering is that it can provide in vast quantities even the rarest of proteins, which might normally be found in only minute amounts within the body. We do not really know what the true therapeutic role of such substances will be until we test them in every conceivable way, a process only possible because of the large amounts of material now available for the first time. Already we can see proteins produced in this way helping patients with cancer, kidney failure and some viral infections.

The future has many exciting possibilities, and we can only highlight a very few. As we have briefly addressed the problem of cancer, let us look at one set of examples of great relevance to the cancer problem. There exists in the body a series of hormones that regulate the growth of other cells. Think of them as part of a very sophisticated regulatory or homeostatic system which. keeps the rate of growth of cells just right for particular circumstances, allowing gradual physical growth during childhood and adolescence, and then an appropriate rate of replacement of worn-out cells in adult life. Many of these hormones controlling growth (or growth factors as they are sometimes called) act only on specific target tissues. For example, there are growth factors that act on the progenitors of the blood cells, and others that influence nerve cells or blood vessel lining cells. Some act only in a local geographic context, influencing the growth of cells right next to them. Most of the hormones are present in blood and tissues in vanishingly small amounts. The purification and detailed study of these hormones have been devilishly tricky and expensive tasks. For example, in order to learn more about growth factors affecting white blood cells, the proteins from twenty thousand mice were purified to get enough of some of these factors for a chemical analysis of just a part of their structure.

Now recombinant DNA technology has changed all that. Over a dozen regulators affecting blood cell growth and development have been cloned in *E. coli* and then prepared in pure form and in large amounts. So have some of the factors responsible for tissue repair after injury. The whole relationship of these factors to the cancer problem is opening up. Are some cancers due to the production of excessive amounts of these factors? Or to a cell switching on the gene for its own growth factor, thereby constantly stimulating itself to more growth? Can cancers be treated by drugs or antibodies that stop the growth factors from getting to their target at the cell surface?

Although complete solutions to these questions will take many years of exacting basic research, certain answers are becoming clear. Scientists working on the study of cancer have shown that sometimes introducing pieces of DNA into a cell will make that cell cancerous. The piece of introduced DNA can be a normal component of the target cell but during introduction (which is a random process) it has not ended up in its usual location in the cell's DNA. As we have mentioned earlier, it is the surrounding regions of DNA that control when a gene is turned on or off, so the introduced DNA in its unusual site may be switched on at the wrong time. When scientists examined the particular introduced DNA (termed **oncogenes** or cancer genes) in cells that have become cancerous they made a remarkable discovery. Frequently, the introduced DNA coded for one or another known growth factors. In other words, the cell would start to divide because it was receiving instructions from the growth factor it was making itself, to continue to grow. The normal processes that turn off growth factor production when the cell has divided would no longer work on the unusually sited gene and the cells multiplied without end, thus forming a tumour. The excitement of this discovery was intensified by results from the detailed study of a group of viruses (called **retroviruses**) that are capable of causing cancer in animals. When infecting a cell, these viruses insert copies of themselves randomly into the host DNA. In some of these viruses there are found DNA sequences corresponding to growth factors or growth factor receptors. (Receptors are the proteins that a cell inserts in its outer membrane. These combine with growth factor present in the circulation and then signal the cell that growth factor is present and it should start to grow). Thus retrovirus infection could lead to a growth factor gene being inserted into DNA in a position where it would be switched on. Again the cell would be fooled into believing it was receiving instructions to grow and would start to divide in an uncontrolled fashion. Thus in some ways tumour cells can be thought of as cells still responsive to normal stimuli which are dividing because of inappropriate instructions.

Growth is frequently linked with a cellular process termed **differentiation**. This simply means the development of some specialized functional capacity, with the end result being a non-dividing cell superbly designed to perform one detailed task. Differentiation, too, is controlled by hormones. Could an undifferentiated, growing cancer cell be treated with purified differentiation hormone, to force it to behave in a more orderly manner and, in fact, to give progeny that are of some use to the body? In experimental animal systems we can give a positive answer to this question, but only more work will reveal its relevance to humans. Genetic engineering has already given us the large amounts of factor we need. We now examine some growth factors that have recently become available for use on patients, with extraordinary results.

Growth factors for blood cells: erythropoietin and colony stimulating factors

The blood found within the arteries and veins of the body is composed of cells and a number of different proteins. There are red cells for carrying oxygen to the tissues of the body and a number of different white cells that are responsible for destroying invading microbes and cancerous cells. The blood proteins are involved in many different activities that include transporting iron and fat around the body, the formation of blood clots and their removal and the retention of fluid within the blood vessels. The blood cells are formed within the bone marrow and it is here that growth factors regulate the orderly development of the cells in response to the body's need. Each particular growth factor acts on a limited number of target cells stimulating them to divide and make new blood cells, thus there are growth factors for red cells and different ones for white cells.

Erythropoietin (EPO) is a growth factor that acts to make new red blood cells. EPO is produced in the kidneys and is released into the bloodstream. It then goes to the bone marrow where it stimulates the early red cell precursors to divide and produce mature functioning red cells. The gene for EPO has been identified and EPO can now be made by genetic engineering. Interestingly, instead of being made in bacteria EPO is synthesized inside a continuous cell line derived from Chinese hamsters. The reason for this is that these cells make a version of EPO much closer in structure to the human EPO than can be made in bacteria. Many human proteins have sugar molecules stuck to them (a process called glycosylation). Proteins made in bacteria lack these sugars so EPO made in bacteria is sugarless. When injected into humans this sugarless EPO is rapidly removed from the blood and destroyed. EPO made in hamster cells (with the sugars added) circulates for many hours and has time to stimulate the early red cells in the marrow. In patients with severe kidney disease the amount of EPO made in the kidneys drops and becomes insufficient to stimulate the formation of new red blood cells. The patients become anaemic and require transfusions. In the USA about one quarter of the 80 000 patients having dialysis for kidney failure need transfusions regularly, consuming about half a million units of blood yearly. We now know that giving these patients EPO will enable them to make their own blood cells again. Transfusion-dependent patients with kidney failure are being given recombinant EPO regularly, with the result that they produce red cells more efficiently and no longer require transfusions. The other effects of this drug are that the patients start to feel very much better in themselves, and their energy and enjoyment of life is boosted dramatically. EPO has been found to be almost completely without toxic side effects. The only requirement seems to be that patients require a dietary supplement of iron because of all the red cells they are now producing. The great thing is that there is now no shortage of this drug and all patients who need it can receive it, that is of course if they can afford it, an issue we will return to later in this book. It is no wonder then that EPO has become a much sought after drug. Marketed under the trade name of Epogen, this product has sales exceeding US$2 billion a year. Patients with a number of other chronic diseases, such as rheumatoid arthritis and liver disease, often have anaemia as well and it seems that EPO is having a major beneficial effect on these patients both in terms of their anaemia and often by improving the disease itself. Another use of EPO is for patients undergoing elective surgery who will require transfusions. These patients can be given EPO a month or two before surgery to stimulate new red cells. The newly produced blood is collected as a routine blood donation. At the time of surgery patients are given back their own blood when they require a transfusion. This ensures a completely safe transfusion and eliminates the risk of contracting AIDS or hepatitis from the use of donated blood. EPO thus is one of those drugs whose development has occurred entirely within a company environment. Amgen, the company that produces EPO and other growth factors for blood cells, started with one employee in 1981 and a capital inflow of US$19 million dollars. By 1999 the company had over five and a half thousand employees and a market value of US$42 billion, a value similar to major companies such as Boeing Aircraft Corporation and Compaq Computers.

In much the same way as happens with EPO, white cell growth factors stimulate early white cells to divide and produce large numbers of mature white cells. These factors, known as **colony stimulating factors**, were discovered and have been extensively characterized by Donald Metcalf and colleagues at the Walter and Eliza Hall Institute and affiliated institutions in Melbourne. There are many situations where a burst of production of white cells is beneficial as a means of boosting immunity. For example, patients with cancer are often treated with very toxic drugs to kill cancer cells. A side effect of the drugs is to kill other dividing cells in the body including white blood cells. Lowering the numbers of white cells in this way makes patients vulnerable to infection. If patients receive a white cell growth factor such as G-CSF (granulocyte colony stimulating factor) then white cells are rapidly produced and the time in which they are susceptible to infection decreases. The upshot of this is that cancer patients could be safely treated with much higher doses of anti-cancer drugs, leading to increased killing of cancer cells but fewer side effects on the patient's bone marrow. G-CSF is commercially available from Amgen under the name Neupogen, and the company sells about US$1.1 billion of product yearly. Another related white cell growth factor GM-CSF (granulocyte-macrophage colony stimulating factor) has been used to increase white blood cell numbers in patients suffering from AIDS. It is anticipated that patients with a variety of different illnesses could be helped to fight infections by treatment with white cell growth factors. In particular patients with burns or traumatic injuries, patients with organ transplants and diabetics could all be expected to benefit from such treatment.

Outside the bone marrow, white cells signal each other to grow and differentiate using an array of messenger proteins called interleukins. The use of one of these interleukins, IL-2 (interleukin-2), was pioneered by Steven Rosenberg and colleagues at the National Cancer Institute in Washington DC. Patients with advanced, untreatable tumours were either given IL-2 directly, or their white cells taken, treated with IL-2 in the test tube and returned to the patient together with more IL-2. There were improvements in some patients with destruction of the tumour and increases in survival times. However, the IL-2 treatment was associated with quite severe side effects such as fever and retention of fluid in the lungs and the experimental treatment is very expensive. IL-2 has now been tried in many cancers often with positive results. Researchers are still learning how to use IL-2 safely and assessing its usefulness in cancer therapy. It is clear that it will be used in combination with conventional therapies and with other white cell hormones such as α-interferon and tumour necrosis factor that are being tested for anti-cancer properties. Another exciting possibility is that IL-2 given after anti-viral chemotherapy may dramatically lower the virus count in HIV/AIDS.

Improving survival after heart attack

The twin big killers in our society are cancer and heart disease. We have discussed how genetic engineering is offering new insights into cancer. It is also showing itself able to improve the outlook for patients with heart disease. Coronary heart disease is the major cause of death in our society. Deposition of a fatty-like material on the coronary arteries, the arteries that supply the heart muscle with blood, eventually leads to blockage. Often a blockage occurs suddenly as a blood clot forms over the fatty material (called atheroma) and the heart muscle supplied by the blocked artery is deprived of oxygen. The patient has had a heart attack (myocardial infarct). If the blood supply is not quickly restored the heart muscle dies, and should the patient survive, a fibrous tissue replaces the muscle. This fibrous tissue cannot contract like muscle and so the heart as a whole pumps the blood less effectively.

Emergency treatment of heart attack involves preventing abnormal heart rhythms and trying to reverse the blockage by dissolving the blood clot. A number of different methods have been used to try and dissolve the blood clots based on injecting molecules that break down the proteins of the clot. One of the more effective of the clot destroyers is a protein called tissue plasminogen activator

(TPA). TPA is an enzyme that is normally present in the blood-stream in minute quantities and is involved in preventing blood clots forming when they are not required. By standard genetic engineering techniques, workers at the US company Genentech made large amounts of human TPA and showed that TPA, injected into patients who had suffered heart attacks, quickly and effectively removed the blood clot, leading to restoration of blood supply to the heart muscle. They showed that once the clot was removed the heart contracted more strongly, improving the flow of blood around the body. It was very important that the injection be given as soon as possible as the longer the heart was deprived of oxygen, the larger the amount of damage. TPA is now available commercially as the product Activase, and is used for treatment of early heart attack and stroke to quickly restore the blood supply to the affected organ.

The future of biologicals

The production of precious proteins through genetic engineering has fuelled the DNA industry until now. In reviewing the choices we have made to illustrate this subject, we are only too conscious of their arbitrary nature. The overwhelming impression to be gained at present is that medical research into biologicals of therapeutic value has entered an entirely new era with the advent of genetic engineering. Things we had not dared to dream of a decade ago now seem conceptually trivial. Just as well the young do not share our inhibitions! They are propelling medicine into an era the limits of which we do not even begin to comprehend.

This chapter has been written as if the boundary between drugs and biologicals were quite sharp, but the genetic engineering revolution will soon disprove this. It is leading to a very precise knowledge of the structure of biologicals, and of the molecules on cells with which they interact. Armed with this new knowledge, organic chemists interested in drug design are making new drugs in shapes that imitate the shape of the most important part of the biological preparation. Such drugs can mimic the desired therapeutic effect. So a new era of pharmacology has also begun.

This chapter has described the process of turning scientific discovery into effective medicines for people. In passing we have mentioned the large profits that are now being made from the biologics in human health. Such large profits are made because the medicines being produced are priced very highly. This is a phenomenon familiar to all observers of the pharmaceutical industry. The companies are commercial entities and as such are concerned with the profitable performance of their business and the return to investors of handsome dividends and increases in stock prices. Those that are unable to do so will not long continue to exist, as they will either be taken over by more financially robust competitors or have their stock sold off to low values and eventually disappear. We have become very familiar with the operation of the implacable rules of the market. Yet we should also remember that the products we talk about here are substances found in all humans. For a period of time, the instructions to make these belong not to all of humanity, but are owned by particular companies. This state of affairs has disturbed many people who feel that aspects of life such as genes and organisms should not be able to be owned, but are the birthright of all. This is a difficult area, because if we are to benefit from discoveries in the form of new medicines or better agricultural products, someone must be prepared to attempt to make a new product, with all the attendant risks. These costs are not small, with the average new drug requiring $500 million or more in development costs. Who will take such risks without some assurance that a profit awaits success at end? However, the prices that are charged for these new products often put them out of the reach of large segments of humanity. How can we balance the desire for innovation and rapid availability of new medicines and equitable access for all? These are not easy problems and as yet we do not have satisfactory answers for many of these. One point to keep in mind is that patents grant exclusivity of manufacture for a period of twenty years, so that the first fruits of the genetic engineering revolution will soon be out of patent. Thus for example, recombinant erythropoietin will be out of patent in 2003. At this time, additional companies, including some sited in the developing world, could commence manufacture. It is hoped that competition will lead to a lowering of the cost of this product and a consequent increase in availability. Many more drugs will come out of patent in the next decade, and it is to be hoped that they too will become available at a lower price so that many more can share in the benefits of the genetic engineering revolution.

6

Genes as diagnostic probes

Bacteria as factories for precious proteins have received such widespread publicity that many readers will already have known much of what we described in Chapter 5. It is now time to move into less familiar waters, in fact into uncharted depths where the immense power of the new technology raises perplexing questions never before faced by humanity. This journey will illustrate how basic research, pursued initially for no purpose other than a search for truth, can lead to profoundly practical end results. Moreover, there is a fair chance that the next four chapters will err on the side of conservatism. They will describe what scientists can do, and think they might one day be able to do, with genetic engineering. Such an analysis can make no allowance for the entirely novel and unexpected results which will certainly turn up on the way, and which could alter the course of research quite drastically. This is what frightens the critics of science, but no formula has yet been devised that can predict which lines of research are 'safe'. It may well be that the new knowledge will be so powerful as to change the established order of things, and will place in humanity's hands the most awesome powers. These are issues which will occupy us for the remainder of this book.

The nature of the diagnostic process

All diseases manifest themselves in ways that give a clue to their nature. Influenza means a fever, a headache, a feeling of malaise, a cough and odd aches and pains. A heart attack usually begins with a severe pain in the chest. Appendicitis shows itself through stomach pains and tenderness in the right lower quadrant of the abdomen. By tradition, doctors divide these natural manifestations into two components, the history and the physical signs. The history is simply what the patient (or the relatives) tell you: 'Doctor, I have a severe pain in the belly'. The physical signs are what the doctor finds out about the manifestations of the disease by a simple bedside examination: if the patient's abdominal muscles are very taut, and he/she practically jumps off the bed if prodded in the lower right portion of the abdomen, appendicitis is on the cards. It is amazing how accurately many cases can be diagnosed by taking a careful history and performing a thorough physical examination. It is also deeply distressing to see instances where the diagnosis of a serious illness has been delayed because of inattention to these two critical components of medical practice. However, and to an increasing extent, such a *clinical* diagnosis is nowadays regarded as only provisional, until more specialized scientific tests are performed. This is not only because an increasingly knowledgeable public wants scientific proof. It is also because most of the clinical rules have exceptions. Loads of heart attacks are 'silent' with no pain at all; some people get the 'flu but do not feel particularly sick, and an appendix can flop into the pelvis giving different symptoms and signs. So, medical science has evolved a panoply of test procedures, biochemical, radiological, pathological, microbiological and so forth, aimed at establishing as objectively as possible the true nature of a disease and also its severity. And we are now moving into the most profound and, in some ways, terrifying phase yet seen—the establishment of the genetic basis of disease.

The genes—those little packages of DNA inherited when one particular sperm meets one particular egg to create a new individual—always represent a mixture of the half coming from the father and the other half coming from the mother. Moreover, because of a peculiar and profoundly important special form of cell division called **meiosis**, each sperm in a man (or each ovum in a woman) is slightly different from every other one of that same individual. Take the case of sperms being generated in the testis. These come from ancestral or precursor cells bearing a full twenty-three chromosomes from that man's father and a further twenty-three from the mother, therefore forty-six chromosomes in all. During meiosis, one of these precursor cells *doubles* its amount of DNA but actually gives rise not to two, but to four, progeny cells.

The sperm therefore has only twenty-three chromosomes, known as a **haploid** set, as opposed to the forty-six chromosomes, the **diploid** set, of every other cell in that man's body. Another and even more curious thing happens during meiosis. Before the two sets of chromosomes go their separate ways, they split and join up at multiple places, recombining that mother's and father's traits into new patterns. This happens differently in each individual meiosis, so that each resulting cell has a slightly different combination, thereby creating great diversity within the sperm population. When a particular sperm from a given father meets a particular ovum from a given mother, a unique individual results, genetically different in all sorts of ways from his or her brothers and sisters.

Another process that introduces differences into the DNA of different people is mutation. Each time the cell divides the DNA must be copied faithfully into a new strand. If an error occurs then one of the daughter strands will have a different base: a **mutation**, and all cells derived from it will share the mutation. If the cell with the mutated DNA is a sperm that fertilizes an egg cell then the resultant person would have that mutation and pass it on to some of his or her children. Now mutations are not a particularly common event, as there have evolved accurate proof-reading mechanisms within the cell that keep the DNA sequence unchanged. This is important, as wholesale changes from mutation occurring frequently would quickly result in a cell becoming non-viable. Mutations can occur anywhere in DNA (although there are some regions of DNA where they occur much more frequently) and if they occur outside of genes they are usually harmless. In fact, mutations rarely produce a favourable change, at least in people. Nevertheless occasional genetic change must occur if living organisms are to be capable of responding to changes in the environment. So, these proof-reading mechanisms sometimes do fail either spontaneously or because environmental events such as cosmic radiation or exposure to some chemicals have damaged the DNA and the repair mechanism has made a mistake in trying to fix the damage. Then that mutation will be passed on from generation to generation. In another family mutation may occur in a different place, again being passed down from generation to generation. Should individuals from these two families produce children together some of these mutations will be passed on, depending on the random process of meiotic recombination we mentioned above, leading to a complex mosaic of mutations in the children.

We now know that in fact mutations are not the only changes that can occur in DNA. It is possible to gain or lose stretches of DNA between generations and there are regions of the DNA where simple repeat sequences can be found. These repeat sequences, often as simple as two bases such as CA, occur in runs of a few to hundreds of copies. During DNA replication, the number of copies of these repeat sequences can change, either increasing or decreasing. The net effect of these mutations, insertions, deletions and changes in repeat sequences is that not all people have the same DNA sequences. These processes operating over millions of years have introduced changes into the DNA sequence which together with recombination, means that no two people (other than identical twins who develop from the same egg–sperm combination) have exactly the same DNA sequence. It also means that particular features, whether mutations or other changes, may be found in some families and not in others. If one considers the human population as whole, then probably there is a variable position in the DNA sequence every 300 to 500 bases.

Previously, we could categorize such genetic differences in a general kind of way. Now, through genetic engineering, we can examine a person's individual genes with great scientific accuracy. This is best illustrated by a few examples. We will see in these examples a rapid evolution in sophistication of method so that our first faltering attempts to look at a single mutation in one gene has been replaced by the ability to scan through the genome for multiple changes.

The haemoglobinopathies—genetic detection at work

In 1904 James Herrick described a peculiar and very serious form of anaemia in which the red blood cells, normally rounded discs, turn into a sickle shape. In 1949 the great chemist Linus Pauling and his colleagues discovered that the oxygen-carrying protein, haemoglobin, was chemically different from normal in patients with sickle cell anaemia. This was the first clear-cut example of a molecular disease. Later research showed that the haemoglobin differed in only one amino acid from normal haemoglobin; a valine had been inserted instead of a glutamic acid. This change profoundly reduces the solubility of haemoglobin after it has discharged its oxygen. Therefore, the haemoglobin comes out of solution and forms a hard crystal inside the red cell. This causes the red cell to assume the peculiar sickle-like shape, and often the sickled cell, lacking the capacity to bend and twist of a normal fluid red cell, cannot get through the finest blood vessels, which in turn become blocked. One single copying error or mutation in the DNA of the haemoglobin gene, one base incorrectly inserted, did the damage and changed the codon for valine into one for glutamic acid. That little error condemned thousands to death! In fact, sickle cell anaemia is by no means rare, and we must ask why the mutant gene reached such a high frequency in the population. It turns out that one needs a double dose of the sickle haemoglobin gene to get the disease, one inherited from each parent. If a person has one sickle gene and one normal gene, the person is a carrier of the trait but is not affected.

In carriers, one half of the haemoglobin in the red cell is normal, and half abnormal. This is not a sufficient change to allow the red cell to sickle or to damage it gravely, but it does change the cell quite a bit. It turns out that the parasites of the worst form of malaria cannot multiply in such changed red cells, so carrier status confers protection from malaria. Obviously, carriers will not be wiped out by the disease and so reproduce and pass the gene onto the next generation. The malaria parasite was so important in some regions, for example parts of Africa, that up to 20 per cent of people are carriers of the sickle cell trait! The horrible price paid is that when two carriers marry, one child in four from the marriage has the double dose, and thus sickle cell anaemia. This genetic burden was passed on to the American blacks when they moved from Africa to America, so they are now left with the disease problem, but with no counterbalancing benefit as malaria does not exist in the USA.

The discovery of mutant haemoglobins, and the diseases they cause, the **haemoglobinopathies**, was important in illuminating one of the great truths of Darwinian evolution, namely that a mutation can be both beneficial and harmful depending on circumstances. The diseases give a profound and direct insight into the relationship between a simple genetic change and a severe illness. Since then, over a hundred mutant haemoglobins have been discovered. So have diseases where the abnormality of the haemoglobin is somewhat more complex than just a simple amino acid substitution. Genetic engineering has been essential in pinpointing the nature of the haemoglobin abnormality in many of these cases. The techniques for the detailed analysis of amino acid sequence of proteins are laborious and very specialized. If you go straight to the gene, or sometimes to the cDNA made from haemoglobin-synthesizing cells (see Chapter 3), you can frequently get to the exact root of the problem much more quickly. Also, if you know exactly what you are looking for, you can sometimes arrange circumstances so that simple and sensitive tests can diagnose a particular disease. Consider the following example. You have a hot probe (see Chapter 3) for a given gene, say the gene for that chain of haemoglobin that goes wrong in sickle cell anaemia. You choose a restriction endonuclease that cuts the gene in the part where the relevant valine (which is changed to a glutamic acid in the sickle haemoglobin) falls. The gene for normal haemoglobin gets split by the endonuclease, but the gene coding for sickle haemoglobin does not, as one of the bases is now wrong for the endonuclease to act. So, when a DNA digest from a few cells of a person is made using that endonuclease, and is then run out on a typical size-sieving gel, the normal person's DNA will have two bits capable of lighting up with the probe, but the sickle cell patient's, only one (Figure 9). This stunningly simple test requires very little DNA, nor do the cells concerned have to have anything to do with blood—the mutation will be in *every* cell, whether the gene is expressed in that cell or not. This type of analysis (called Southern blotting after Ed Southern, the scientist who developed it) can be extended considerably by the use of different enzymes to digest the DNA at different sites and by the use of different hot probes.

With the increasing sophistication of genetic techniques, we would now consider this methodology to be quaint and somewhat archaic. Newer, faster and more accurate methods have been developed that do not require the fortuitous creation or loss of a restriction site as a result of the mutation. These methods examine the exact DNA sequence itself, and have largely replaced the method described above. It is instructive to realize that this method lasted less than ten years before it was replaced by improved methods.

Let us now examine the practical consequences of such a technique.

Hot probes and foetal diagnosis

Until recently antenatal diagnosis of sickle cell anaemia, thalassaemia and related diseases was carried out by specialized tests on foetal blood. As there are over two hundred million carriers for the inherited disorders of haemoglobin, and about two to three hundred thousand severely affected children born into the world each year, and as there is no definitive cure for any of these diseases, the antenatal diagnosis is by no means an academic matter. In practice, it has not been possible to dissuade known carriers from marrying other known carriers. Indeed, most physicians and scientists who have worked in this difficult field have little sympathy for the brave new world in which the priest would say: 'Do you, John, a carrier of sickle cell trait, take this woman, Mary, a guaranteed non-carrier, for your lawful wedded wife?' Not only does it appear impossible to change mating patterns through genetic counselling, but also experience has shown that a significant number of carriers misinterpret the information received at counselling, and are left with the belief that they themselves have a serious disease. Based on those screening programmes for sickle cell carriers that have been carried out, it seems preferable to offer couples the possibility of accurate antenatal diagnosis, and the choice of a termination of pregnancy in the one case in four where disease is actually present in the **foetus**. We are aware that therapeutic abortion is abhorrent to many people. However, the public health problem is so great, particularly in some developing countries, that the option of allowing these hundreds of thousands of mortally ill children to be born and to drain the scarce health resources of their nations as they eke out a miserable existence of a few years' duration is even more abhorrent.

The problem with waiting for a foetal blood sample is that it can only be reliably obtained at about eighteen weeks gestation, that is to say when the mother is four and a half months pregnant. This means an agonizing wait for the mother—fourteen weeks from the first missed menstrual period—and a difficult type of termination (frequently by a Caesarian-section-like surgical operation) as

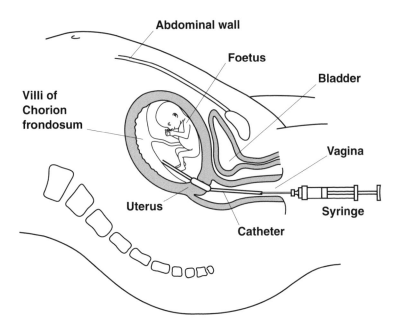

Figure 9. Diagnosis of sickle cell anaemia by restriction endonuclease mapping of the globin gene.

DNA from the foetus is first sampled by inserting a needle into the mother's womb and extracting a small piece of the lining of the placenta, a part made up of tissue from the foetus. DNA is extracted from this tissue for analysis. In this disease, the gene for one of the globin chains of the haemoglobin molecule has mutated so that a valine is replaced by a glutamic acid. Some cells from a developing embryo are obtained, and the DNA is extracted. It is treated with a restriction endonuclease enzyme capable of recognizing the site where the valine in question falls. The resultant DNA digest is placed on to a size-sieving gel which is then treated with a 'hot probe' for the globin gene. An X-ray is taken, and if the haemoglobin is normal, it will be split by the endonuclease, and two different fragments of DNA will each contain a piece of the globin gene. The hot probe hybridizes to each of these, so two pieces of DNA yield dark bands on the X-ray. If the haemoglobin is sickle cell haemoglobin, the endonuclease does not act, so only one dark bank is present on the X-ray, corresponding to the one piece of DNA containing the mutated globin gene.

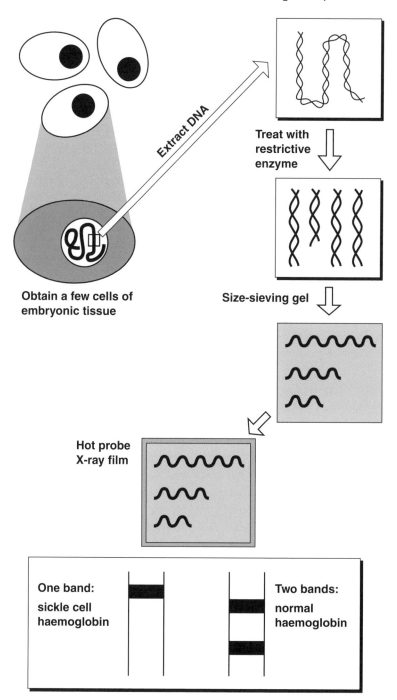

Extract DNA

Obtain a few cells of
embryonic tissue

Treat with
restrictive
enzyme

Size-sieving gel

Hot probe
X-ray film

One band:

sickle cell
haemoglobin

Two bands:

normal
haemoglobin

the child is quite well formed at that time. A newer technique is now available. You really only need a tiny amount of DNA to make the diagnosis if you know what you are looking for and you have the right hot probe. Therefore it is now possible to sample foetal genes as early as two weeks after the first missed period, and quite reliably four to six weeks after (eight to ten weeks gestation). The technique is known as trans-cervical aspiration of **chorionic villi**, but the formidable name need not obscure the essential simplicity of the method. Remember that genes are *present* in cells even when they are not *expressed* in them. So, the haemoglobin genes are present and expressed in red blood cell precursors; they are present but not expressed in all the other cells of the body. Inside the uterus the developing **embryo** is surrounded by a layer of tissue called the chorion, which thrusts slender, finger-like projections known as villi into the uterine wall. These chorionic villi are destined to form part of the foetal contribution to the placenta (or afterbirth), which also has a maternal component. At the early, delicate stage of pregnancy, a few chorionic villi can be aspirated via a long thin tube that is inserted into the vagina, threaded through the cervix, and positioned by the obstetrician with the use of an ultrasound monitor. The procedure is painless and less distressing than **amniocentesis** which involves a puncture wound in the skin. The procedure increases slightly the risk of miscarriage, but is quite safe for the mother. When serious disease is diagnosed as early as this, the mental and emotional trauma of a therapeutic termination by a simple curette is certainly less than that by open operation after the baby has quickened in the womb.

With the new gene detection methods, it is possible, at least in theory, to take the material provided by aspiration of chorionic villi and read the DNA sequence of the gene of interest. There are over 100 inherited diseases for which the gene involved has been identified. Not all of them are as simple as sickle cell disease in which a single change in one amino acid is responsible. For many diseases, such as cystic fibrosis and muscular dystrophy, the genes are very large and there are many mutations and other changes associated with the disease. Careful examination of cystic fibrosis patients has shown that there are more than 850 mutations, deletions and insertions that lead to altered function of the protein responsible for the disease. It would be very difficult and expensive to examine every base in the sequence and check that it is correct. In fact, the genetic tests that are available look for the most common changes such as the loss of the amino acid phenylalanine at position 508 in the responsible protein (the cystic fibrosis transmembrane conductance regulator or CFTR). By scanning for 10 or 20 of the most common abnormalities, these tests can identify perhaps 90% or more people carrying defective genes. Often such tests will work better in one ethnic group than another, because they have been developed by studying one group, such as Caucasians living in the USA. As we have discussed, particular mutations are passed down in families and are likely to be identical in groups that have a pattern of marriage within the group. Other mutations arising spontaneously in other populations will need to be discovered and specific tests developed.

Gene linkage—unlocking the black box

The defect in the gene that causes sickle cell disease was well known, and consequently, it was the first genetic disease for which a specific genetic test to identify carriers and affected individuals could be developed. Many other common genetic diseases were recognized but in almost all cases, it was not known which protein or proteins were malfunctioning. Thus it was not possible to examine the genes responsible. In just a few short years however, we have been able to identify the causes of many severe and common diseases. The causes of diseases such as cystic fibrosis, one the commonest inherited diseases in the world, Huntington's Chorea, an untreatable dementing disease of adults, and many of the muscle diseases such as Duchenne's muscular dystrophy, have all now been found. We know what gene is abnormal and consequently what protein is not performing its functions properly. We do not yet understand every single feature of how some of these abnormal proteins cause disease, but we know enough to devise sensitive diagnostic tests. How did these advances come about so rapidly, when we understood so little about the genes involved? The answer is a set of gene mapping techniques known as linkage analysis. We mentioned this in Chapter 4, when we discussed the genome project and the aid it will give to mapping projects. Linkage analysis requires us to make no assumptions about a disease, other that it be inherited. We then start to examine a family in which the disease occurs. In the initial stages it is very helpful to examine a large family in which there are several cases of the disease. We have mentioned how during sexual fertilization the newly formed egg receives one set of chromosomes from the father and one set of chromosomes from the mother. Each set of chromosomes has gone through a recombination process which has shuffled bits of DNA from the father and the mother into somewhat different combinations. What is the chance that two particular pieces of DNA will be inherited together? It turns out that if these pieces of DNA are close together on the same chromosome they are likely to be present in the same sperm or egg after the shuffling process. The further apart they are, the more likely they might separate, and if they are on separate chromosomes they are inherited independently.

Closely situated genes that are inherited together are known as linkage groups. If we have a means of finding one gene then the odds are high that other genes in the linkage group are also present. We can find whether a person has a particular gene by the technique of Southern blotting already mentioned. Now comes the twist that makes this technique so powerful for detecting genetic defects. To find a damaged gene we do not need to know it precise location: to start with, we need only find a piece of DNA in the same linkage group. To do this we can take a number of hot probes that we know come from different parts of every chromosome in the genome and look at the individuals in the affected family. Remember the hot probe gene need have nothing to do with the disease. It only has to be close enough to the disease causing DNA to serve as a useful marker for it. Can we find a hot probe that gives a particular pattern in individuals with the disease but not in healthy people? If you look with enough probes you are bound to find one that is linked with disease. Many scientists were involved in the vast collaborative enterprise of compiling a set of random probes. For a while it became an area of intense competition and of entrepreneurial activity as companies tried to find probes linked to important genetic diseases, keeping the identity of these probes secret. The success of the human genome sequence which has been made freely available to all, provides the means to identify such markers and has fortunately brought to a close this period of secrecy.

Once a probe with linkage had been found, then standard genetic engineering techniques and a good deal of hard work enabled scientists to crawl along the chromosome until they found the diseased gene itself. So without knowing anything about how the illness was caused we have zeroed in on the gene at fault. Gene probes to linkage groups can be used as diagnostic tests to predict whether a foetus will inherit the disease, in exactly the same sort of test as described for sickle cell disease. Initial tests tend to be useful only in particular families, as linkage groups vary in genetically distinct groups of people. However, as a probe moves closer to the damaged gene it becomes more generally applicable as a diagnostic tool.

But the story of linkage groups does not end there. It can be extended to diseases where **heredity** may play only a partial role. We know for example that many factors predispose to heart disease. High blood pressure, high cholesterol levels in the blood, smoking and a poor family history are all risk factors for heart attacks. A family history suggests there is some genetic factor involved. By linkage analysis we could home in on part of the DNA that carries this poor family history. We could in theory predict whether someone does or does not have this predisposition to disease. There are a number of diseases with an inherited component including manic depressive illnesses, schizophrenia and autoimmune diseases such as rheumatoid arthritis, where these techniques could also be applied. However, to perform such linkage analysis, we stand in need of a good pedigree, where the disease runs in a group of related individuals. It becomes much more complicated if the group being studied is made up of unrelated individuals, where the gene defects may be different and there are many other changes unrelated to the disease. An Icelandic scientist working at Harvard University, Dr Kari Stefansson, realized that the people of Iceland were a relatively homogeneous group whose pedigree was known for many generations and whose medical conditions were also well documented. They thus represented an ideal study population of 270 000 for performing linkage analysis for many complex diseases. He formed the deCode Corporation, and in a controversial deal with the Icelandic government, licensed the rights to use the Icelandic population for genetic research. The company focuses on 35 diseases including common diseases such as osteoarthritis and asthma and uses the techniques we have described to find candidate disease-associated genes. The large pharmaceutical company Hoffman-La Roche has signed an agreement with deCode to use the results of its research. The establishment of this company was the occasion for much debate over the rights to privacy and the issues of commercial exploitation of public records. It ultimately required an Act of Icelandic Parliament to resolve the issue, although even that has failed to silence critics.

There are other applications of linkage analysis. Many observers have noted that complex behaviour patterns such as drug addiction and alcoholism run in families. Some suggest that this is due to social circumstances such as poverty, failure of access to counselling and other social services. An alternative view is that this is due in small or large part to genetic factors, and that these genes can be identified by linkage analysis. This is a highly controversial area and its results may turn conventional treatments of these conditions on their heads. Other aspects of the human condition have also been subjected to linkage analysis. A study some years ago described a significant linkage between a region of the genome and homosexuality. How about intelligence or artistic ability? We certainly know of famous families such as Johann Sebastian Bach and his talented children, where these achievements were found in a family group. Do genes controlling such abilities exist and can we find them? In Chapter 7, we will discuss methods for altering genes in humans. Should such methods be used to alter an individual's likelihood to behave in a particular manner or to enhance some of these abilities in others?

For now we will just make a few points about the less controversial subject of identifying genes that predispose to physical illnesses such as heart disease. There are a number of obvious benefits from having such tests. They allow a more reasoned planning of one's life. If you have a predisposition to a lung disease such as emphysema, then it would be unwise to take a job in a dusty environment such as a coal mine. Similarly if you have a predisposition to heart disease then a particularly stressful job as an air traffic controller might make your health worse. The existence of predictive tests of such power could also cause some problems. An insurance company might decide not to offer you life or health insurance or charge you much higher premiums because it doesn't like the look of your tests. It might also refuse to insure you unless you provide it with certain test results. The insurance companies might argue that they are entitled to know such information in order to charge appropriate premiums and protect the financial interests of their other clients. Such problems of course would not be unique to these genetic tests, as we have seen the same battles raging with AIDS testing. Community debate leading to guidelines and ultimately legislation is one pathway by which such problems can be met. Undoubtedly, the subject of genetic profiling will continue to hotly debated. The movie *Gattaca* provides one bleak scenario of what would happen if this technology becomes inappropriately applied. We remain hopeful that a consensus will be reached that will allow the benefits of this technology to be realized with minimal disadvantage to the less advantaged in our society.

Genetic fingerprinting

As described earlier, techniques such as Southern blotting can in fact show up these differences between individuals as changes in the fragment size seen by a particular hot probe. Theoretically, then, if we analysed a person's DNA with enough restriction enzymes and hot probes we could describe his DNA as a list of fragment sizes that would specify him or her uniquely at the genetic level. No other person would have exactly the same fragment patterns.

However, this would be an extraordinarily tedious and expensive way to do it. Dr Alec Jeffreys, a scientist working in England, has pointed out a simple means of examining each person's DNA to highlight unique patterns. His procedure is based on the existence of short sequences that are repeated within the DNA. Unlike genes for proteins that occur usually once or a few times these repeated sequences occur hundreds or thousands of times at different positions in the DNA. Because they are repeated, a hot probe of a repeated sequence lights up many bands on a Southern blot of DNA, giving an appearance somewhat reminiscent of the bar codes on supermarket products.

The variability of position and length of these repeated sequences is such that no two individuals (except identical twins) have the same pattern of bands. Crucially, though, the pattern of bands in an individual is inherited half from the father and half from the mother. Thus every individual can be associated with a unique pattern of bands and family relationships can be documented with a high degree of precision. Enough DNA to perform this analysis can be obtained from a drop of blood, fifteen hair roots or a speck of semen. This technology has already been used in the courts to decide paternity, identify rapists and resolve immigration disputes. For example an Indian immigrant wanted to bring a young boy he claimed was his son into the UK. Immigration authorities challenged the identity of the boy but genetic fingerprinting conclusively proved that the child was indeed the man's son.

Other variable sequences found in **mitochondria** and elsewhere in the genome such as the Y chromosome have of late been used to trace pedigrees and solve various historical issues. For example it has been suggested that Thomas Jefferson fathered a son with a female slave Sally Hemings, because direct male descendants through the Jefferson line and the Hemings line shared the same pattern of sequences on the Y chromosome. Of course such sharing would also be the case if the father was one of Thomas Jefferson's own sons as they too would carry the characteristic Y chromosome. Of almost as much interest as the science itself was the ensuing controversy, including the charge that interpretation of the scientific results had been slanted to give comfort to President Clinton during a time in which his own sexual conduct was under question. By similar PCR methods, patterns detectable in mitochondrial DNA were used to establish that bodies found buried outside Ekaterinaburg were those of members of the Russian royal family killed during the Russian Revolution. It is certain that the precision of these tests will increase with time, and interested readers are referred to the story of the trial of Mark Alan Bogan (http://www.med.monash.edu.au/reshapinglife) to see how far this technology has come.

Polymerase chain reaction—finding the needle in the haystack

The area of genetic engineering is technology driven. New techniques are willingly embraced and pushed to their limits, to advance the level of knowledge and exploit practical spin-offs. The technique of gene amplification by the polymerase chain reaction (PCR) is such an example; this is a technique of great elegance and quite extraordinary power and it has revolutionized many aspects of molecular biology from the most basic to the most applied. Invented by Kary Mullis and developed further by Randy Saiki, Henry Erlich and colleagues at the Cetus Corporation in California, PCR is a method for making many copies of a gene, but unlike cloning in bacteria, PCR is done in the test tube by a chemical process. What formerly may have taken a week can now be accomplished in just a few hours.

The essential principles of PCR are illustrated in Figure 10. We start with a mix of DNA containing a gene we are interested in. This DNA may be extracted from a tissue biopsy, a blood sample or a sample of chorionic villus from the placenta. We mix in many copies of two short stretches of synthetic DNA of a sequence complementary to the gene of interest that are separated by several hundred base pairs. Finally a heat-resistant enzyme that copies the DNA extending our short synthetic DNA pieces along the gene of interest is added to make fresh double-stranded copies. By raising and lowering the temperature we manipulate conditions in the test tube in such a way that the synthetic DNA pieces come into contact with our gene of interest and are extended. We then separate them by raising the temperature and bring them together again by cooling to allow further extension. As shown in the figure, each time this cycle is performed the amount of our target gene is doubled. After ten cycles there are over 1000 copies, after twenty cycles over a million copies, and so on by powers of two. This ability to amplify a gene or part of it, to find a needle in a haystack, has many diagnostic uses including improved diagnosis of genetic defects, as well as uses in forensics. We will look at three further examples of PCR and its utility.

Gene probes for autoimmune disorders

In 1980 the Nobel Prize for Physiology and Medicine was awarded to a trio of scientists who discovered an extremely complex but important group of genes known as the **major histocompatibility complex**. These genes control a person's tissue type, that is, they determine how strongly or weakly a skin or kidney graft from one person will provoke an immune attack in another person. They also determine immune responsiveness to a wide variety of foreign agents. We now have a much more accurate, and rapidly evolving,

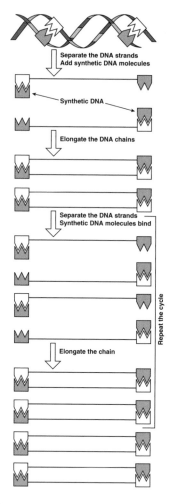

Figure 10. The technique of polymerase chain reaction to amplify DNA.

PCR enables us to make many copies of a particular stretch of DNA. In a test tube we mix together some DNA which can be the total DNA of a human cell. We add short synthetic DNA sequences flanking the region we wish to amplify. These short sequences find their partners in the DNA mix and are then extended to copy the strand. By repeating this procedure many times, usually about thirty, we end up with the specific stretch of DNA flanked by the short synthetic pieces amplified to a large amount. This can be used as a hot probe for diagnosis or to tell us whether the region is normal or mutated.

picture of this genetic region through recombinant DNA technology. It appears that there are no fewer than fifty genes involved altogether, though about ten of these are turning out to be the truly important ones. An exciting area of study is the genetic basis of certain important diseases called autoimmune diseases.

The control of nature's defence system is multifactorial, and the immune regulatory genes of the major histocompatibility complex are only one element in a network of interacting control loops. This complex system sometimes runs amok, and the immune defence cells then make antibodies not, as is their duty, against invading microbes, but against some vital constituent of the body itself. These so-called autoimmune diseases include insulin-dependent diabetes, many forms of anaemia, most forms of thyroid disease, chronic diseases of the liver, kidney and nervous system, and different sorts of arthritis, together representing a large proportion of internal medicine. People with certain major histocompatibility genes run an increased risk of particular autoimmune diseases. This concept of 'relative risk' is unfamiliar to most people. It does not mean that the gene directly causes the disease, but rather than the person is more susceptible to whatever is the true causative agent. One of the problems has been that the exact gene conferring increased risk has not been pinpointed within the broad genetic region of the major histocompatibility complex. Now the genetic engineers can use PCR methodology to amplify particular genes in the complex and read off directly the sequence of the gene. Particular sequences are associated with very greatly elevated risks of contracting disease.

An important practical consequence of tissue types is the requirement to match them for successful transplantation of organs. Until recently, all **tissue typing** was done by using, as the typing reagent, serum from people who had had a transplant or had received multiple blood transfusions, or women who had borne several children. In all three cases, the immune system would have been provoked into making antibodies against various tissue **antigens**, namely, those of the transplant donor, those of the donors of the blood samples, or those of the father(s) of the many children. This typing methodology worked surprisingly well, given the great complexity of this series of genes, but it is being replaced by a direct test of the DNA using pure probes for individual genes. The gene probes can be arrayed as multiple sequences attached to a wafer of silicon, the so-called 'DNA chip', and the position of the amplified DNA from the patient can be read directly off the chip, giving the patient's tissue type accurately and quickly. In this technique, short DNA sequences corresponding to all the possible variants of the gene in question such as the histocompatibility genes are placed on a slice of silicon. DNA from the test subject hybridizes and binds to its own specific complementary sequence. This binding is identified and the tissue type of the patient has been determined. Perhaps the most famous recent use of tissue typing technology by PCR was in the O. J. Simpson trial, where blood at the crime scene was tissue typed and matched to his type, suggesting his presence at the crime scene. As the events played out, the contention became not whether the test identified an actual sample of Simpson's blood, but whether its presence at the crime scene might be because it was planted there by other persons.

Genetic engineering to diagnose viruses gone underground

Different species have sequences that are unique to that species. Thus it is possible to set up PCR tests to examine for the presence of bacteria or viruses in various samples. It is becoming routine now to diagnose infections by PCR using synthetic DNA that will only hybridize to the organism you are seeking. Thus tuberculosis can be diagnosed by taking sputum from a patient and performing a PCR reaction with tuberculosis specific primers. This is very much faster than waiting for traditional methods and patients can be started on appropriate treatment much earlier. Similarly it is becoming a routine method of screening for contamination of food to take samples of the product under test and do a PCR reaction looking for dangerous bacteria, such as the new strain of *E. coli* O157H7. This bacterium has caused a number of deaths due to contamination of meat products such as hamburgers or unpasteurized fruit juices. *E. coli* is a very common organism and is usually quite harmless. It is only when it has picked up an extra piece of DNA that codes for a nasty toxin, that it causes disease. The PCR reaction focuses on that specific piece of DNA and can thus discriminate between dangerous and more benign bacterial contamination. Finally, individual bacteria in a species have repetitive sequences that mark them as an individual in the same way that different humans can be told apart by their microsatellites. Thus it becomes possible to trace the movement of a single bacterial strain from patient to patient. In this way we can trace the spread of tuberculosis through a hospital or the source of an antibiotic resistant organism. Such techniques could also be used to trace the source of release of bioterrorism agents such as anthrax and determine where the anthrax bacteria came from.

We mentioned in Chapter 3 that viruses can jump into and out of the genes of a cell with surprising ease. This is true also for human cells, and when a virus integrates into cellular DNA it usually stops multiplying within that cell, and goes underground, so to speak. Nevertheless, those viral genes can still exert their bad effects on the cell in a variety of ways, for example by causing the cell to make abnormal proteins at the virus's command. Further-more, the virus that has gone underground can sometimes re-emerge with full pathogenic strength. An interesting group of such viruses is the herpes viruses. *Herpes simplex* type 1 causes cold-sores, and *Herpes simplex* type 2 the sores of genital herpes. In between attacks, the virus goes underground and lies dormant until reawakened by some event such as a fever, an emotional shock, sunburn, etc. *Herpes varicella*, the cause of chickenpox, goes to sleep in the nerve ganglia outside the spinal cord, but occasionally comes out decades later in the painful and at times serious condi-tion called shingles or *Herpes zoster*.

PCR gives us a powerful and rapid tool to diagnose the presence of viruses that have integrated themselves into the host cell. With the right sequences of synthetic DNA we can cross-examine cells biopsied from the human body to determine whether a given virus lies dormant in the cell. Previously such a technique could only be done after much laborious and incredibly finicky work on tissue sections.

In AIDS it has become crucially important to identify as early as possible those people who have become infected with the HIV virus. When someone becomes infected with the virus there is a period of between one and six months before AIDS antibodies are detected in that person's blood. It is the presence of such anti-bodies that is used as the screening test by blood banks to exclude AIDS-infected blood from our blood supplies. Thus there is a short period when an infected individual is not recognized as such by conventional serological tests. PCR has helped to remedy this problem, although it is unlikely it can ever be completely solved. PCR is able to detect very low levels of virus, so it is able to identify infected people from very shortly after they get the virus. The test is performed on a sample of blood and is capable of detecting a single AIDS-infected cell among millions of normal blood cells.

Detection of virus is important not only for diagnosis of individual patients but also to give us deeper insight into causative mechanisms of disease. Despite its short history, the field has already shown us some important things related to viruses and human cancer. For example, the virus causing glandular fever, an extremely widespread virus which for most people causes only a mild and transient disease, is present in dormant fashion inside the cells of two forms of human cancer, namely Burkitt's lymphoma and nasopharyngeal carcinoma. How and why most people get over the virus attack while a tiny minority develops this terrible complication is obviously the subject of very intensive research. Another fascinating example is the hepatitis B virus. This is the agent for a severe form of acute hepatitis, but is also involved in some types of chronic liver disease and in liver cell cancer. In both of the latter cases, some individuals become hepatitis B carriers while a few become carriers only of the viral genes. Again, the diagnostic probe for the presence of the virus is proving a valuable tool for just what is going on inside the diseased liver cells.

We will mention one final area in infectious diseases where PCR-based techniques are being used.

Palaeobiology—the truth behind Jurassic Park

The capacity of PCR to amplify a very few DNA molecules into amounts that we can work with has been utilized to examine DNA from ancient sources. Once an organism dies, the components of its cells, including the DNA, start to break down. For DNA this means that the long chains of nucleotides start to break into smaller fragments. Only under quite unusual environmental circumstances will this process be inhibited. However there are a number of circumstances, such as entrapment in peat bogs or in amber, in which the pace of degradation may be slowed or stopped. If DNA fragmentation has not gone into tiny pieces, it is theoretically possible to recover this DNA by the PCR process. We really don't know what the limits might be, but a generally agreed figure is that beyond 100 000 years, the condition of the DNA is too bad for PCR to work, at least using our current methods. The most generally used PCR method amplifies a region of DNA between two synthetic DNA primers (Figure 10). Thus most work has attempted to amplify regions of genes that are conserved among all known life. It is possible to sequence such amplified DNA and compare the differences between that and present day species. This gives us an insight into how rapidly these sequences change over time and helps us interpret the pace of evolutionary change.

Examples of successful recovery include a 379 base pairs fragment of DNA from mitochondria found in a piece of the right thigh bone of a Neanderthal discovered in 1856 near Dusseldorf, Germany. It is thought this bone fragment may be 30 000 to 100 000 years old. Other scientists recovered a 122 base pairs sequence of mitochondrial DNA from the ankle bone of an Egyptian mummy which was probably 2000 years old. Short sequences have been recovered from woolly mammoths and sabre-tooth tigers. However such fragments are minuscule in comparison to the billions of base pairs needed to encode a complete organism. It is not clear how we could ever approach such a goal. The only form of success to date that is somewhat similar is the successful growing of bacterial spores preserved in ambers. Spores are specialized forms of bacteria that are adapted to surviving under harsh conditions. Mammals do not have a similar adaptation. It would seem that Jurassic Park must for the moment remain shuttered and closed.

An intriguing more recent 'fossil' is the influenza virus strain from the great pandemic of 1918 and 1919. This form of flu, called Spanish influenza, killed over 20 million people in the years following the end of World War I, making it the worst infectious pandemic in history. Scientists recovered RNA from that virus from a formalin-fixed, paraffin-embedded sample of lung tissue prepared during the autopsy of a victim. Additional samples were obtained from another victim and one frozen sample obtained by in situ biopsy of the lung of a victim buried in permafrost since 1918. Analysis revealed that this virus is more closely related to human and swine flu strains than to bird strains. However we are still at a loss to understand why it was so fearfully deadly. Many factors determine the outcome of an infection, including the status of the host, and these may well have influenced the death toll at that time.

Once again, the perceptive reader will gather that, by these few examples, we have really only scratched the surface of genes as diagnostic probes, and the uses of PCR. The obviously genetic diseases are, of course, very much in the minority. Yet a great proportion of human disease has *some* genetic component. The real problem has been to get a handle on human genes—to detect them, to classify them, to study their arrangement and interactions. Genetic engineering has given us that handle, and will alter the landscape of diagnostic procedures so that, in twenty years' time, it will be virtually unrecognizable.

7

Gene therapy, cellular engineering and human cloning

Genes can be analysed and studied through recombinant DNA technology. They can be removed, purified and transplanted into cultured cells. But can genetic surgery be performed in a real-life setting? Are there prospects for intervening to alter the genetic constitution of an intact, living human being? This question not only poses a formidable technical challenge, but must surely rank as one of the most profoundly thought-provoking social and ethical issues of our times. To place it in context, one must first ask to what extent genes travel around in nature itself.

Mobile genetic elements

T. S. Kuhn has become famous because of the theories propounded in his book, *The Structure of Scientific Revolutions*. His thesis, essentially, is that there are two sorts of science, 'normal' science and 'revolutionary' science. Normal science is what the great majority of scientists are involved in, day by day, month by month, painstakingly exploring all the ramifications of established leads till a large body of specialized knowledge grows up, creating a pattern or overall framework for looking at particular sets of phenomena. Revolutionary science, on the other hand, is that profound shift in perception, which follows a major discovery that just refuses to fit into the established paradigm. Revolutionary science, Kuhn believes, is born when normal science is in deep travail, because a generalization no longer fits all the observed facts. The shift from Newtonian physics to Einsteinian relativity and to the quantum theory is often cited as the classical example of revolutionary science, with a shift in paradigm.

Kuhn's analysis does not accurately describe biological science, which appears more like a gradual, progressive drive forwards, with deeper layers of understanding not negating, but expanding, previous insights. Nevertheless, there is a deep truth in the view that exceptions to the accepted rules must be taken very seriously. Fundamentally, the role of genetics was built up around the view that the genes were the sacrosanct guardians of a cell's or an individual's potential. These genes were inherited from the parents, and that was that. Very occasionally, a copying error occurred, leading to a **mutation**, and this hit-and-miss process was the source of variation on which Darwinian natural selection could act. When genetics was already a mature science, it was found that the genes were composed of DNA. So the DNA was regarded as the stable, heritable material, which was faithfully copied down the generations. In higher organisms, as already discussed, the DNA is packaged into a **nucleus** in the form of separate double-helical strands which, together with certain other regulatory molecules, constitute the **chromosomes**. In these higher organisms, the complex process of meiosis can scramble maternal and paternal traits, so adding to the pool of diversity on which natural selection can act.

This was the unshakeable dogma, but about thirty years ago it was challenged from a surprising source. Barbara McClintock, an American plant geneticist working with maize, found a mutation that changed the colour of the cobs. This mutation reversed itself surprisingly frequently, and, without going into all the fascinating technical details, it turned out that what was at work in the mutant strain was a **mobile genetic element**, in other words, a jumping gene. This jumping gene could move about from place to place on a chromosome, or it could jump from one chromosome to another. Once you admit that genes can jump about in this bizarre way, it is not such a large step to the suggestion that they could move from one cell to another. There are about thirty to forty such jumping genes in maize, and for opening up this field, which later was to become so essential to genetic engineering, McClintock won the Nobel Prize for Medicine in 1983, the first solo investigator so honoured in twenty-two years. Jumping genes have also been found in the fruit-fly, one of the favorite tools of classical genetics, and in this species it is estimated that as much as 5 to 10 per cent of the

DNA is capable of moving around. We know now from the genome sequencing projects that there is a great deal of movement of genes between bacteria, and this phenomenon dubbed 'lateral transfer' can move many genes simultaneously. In plants and animals, this sort of wholesale transfer of genes has not been found, although mobile elements can jump between species. So we have to accept an overall picture that allows at least a certain amount of mobility of genes as part of nature's plan. The challenge, of course, is to understand that natural process more fully and to harness it to useful purposes.

Mobile genetic elements are being investigated in many centres for their intrinsic interest as a genetic system, and for their potential use in genetic engineering as a novel kind of vector. But they are by no means the only approach being actively explored in order to get genes into cells.

Cancer viruses: how to get bad information into animal cells

So much for bacteria, plants and flies; but can one get new genetic material, let us say a single gene, into the cells of higher animals, including humans? The answer is yes, and extraordinarily enough, the way to do it was first discovered through the study of that dreaded disease, cancer. This section will therefore describe cancer-causing viruses belonging to the group of **retroviruses**, an acronym for the reverse transcriptase enzyme that allows them to do their dirty work. The story represents the most exciting of molecular detective work, and without question the biggest conceptual breakthrough of the last two decades in the cancer field. It is no exaggeration to say that it has cancer research institutes all over the world buzzing with renewed energy as, at long last, we appear to be gaining insights into the causes of this most feared group of diseases.

Cancer is all about the control of cellular growth. It is normal for the cells of the body to divide, but in health the pattern of such division is strictly controlled. Not only must cells divide under strict control but also cells must die at the appropriate time. Skin cells are shed each day, and are replaced by new cells coming from the deeper layers of the skin, so maintaining the status quo. If the skin is stimulated, as by heavy manual work, growth accelerates, the skin thickens and the hands become callused, but to a degree that is strictly in accord with certain regulatory rules. If the skin is accidentally cut, a different pattern of cellular growth immediately begins to bridge the gap. This new growth stops when the cut is healed. Cancer is an extreme form of breakdown of the regulatory processes limiting growth, so that cellular division goes on with progressively fewer and fewer checks and balances, until the body is destroyed by wildly proliferating malignant cells. The cells may either divide when it is inappropriate, or fail to die when signalled to do so.

It has been known for many years that certain curious viruses can cause cancers in experimental laboratory animals, though under circumstances that are frequently highly artificial. More recently, it has been possible to use these viruses to create cancer artificially in the test tube. Cells are first permitted to grow normally and peacefully, under good control, in glass dishes. This is the technique called tissue culture, in which the cells are housed in a moist, controlled, warm atmosphere and are fed a growth medium having most of the constituents of the fluid that normally bathes the tissues. Normal cells grow until the bottom of the dish is just covered, and then they stop, just like the cells healing a cut stop when the gap is filled. When a cancer virus is added to the culture, the cells infected by the virus continue to grow and pile up on top of one another in a rather disorganized fashion. They have been transformed into cancer cells. While it is unlikely that many human cancers are caused by such transforming retroviruses, this model system has been widely studied by cancer researchers because of what it can teach us about the basis of the cancer process.

How viruses cause cancer was a mystery until recombinant DNA technology produced a startling new set of biochemical answers. First, it was found that many of these viruses had genes made *not* of DNA but of RNA. However, the virus forces the cell it infects to make the enzyme reverse transcriptase, which copies the virus RNA sequence faithfully back into DNA. Next it was shown that this copied DNA could integrate firmly into the DNA of the cell itself. The virus had foisted new genes onto the cell! This led to the next major clue, namely that some of these genes were cancer-causing genes, causing the production of proteins which sent the cell wild. Such cancer-provoking genes are called oncogenes, and so far more than sixty different ones have been identified, each carried by a different virus. Oncogenes can now be prepared in pure form through genetic engineering technology. These pure bits of DNA can be artificially injected into normal cells and, without any virus involvement, transform the cells into cancer cells. So, in a real sense, oncogenes cause cancer.

How might a gene turn a cell into a maverick exhibiting uncontrolled growth? As discussed in Chapter 2, genes contain coded information for the synthesis of proteins. It was clearly important, therefore, to find out what sorts of proteins would be made when an oncogene was switched on. A major breakthrough was the work of Dr Ray Erikson, then at the University of Colorado in Denver. He proved that one of the best-studied oncogenes codes for an enzyme known as a **protein kinase**. In the cell, we find little bundles of molecular energy termed ATP that provide the power for living proteins (Appendix B). This protein kinase catalyses a reaction which snatches phosphate from ATP and places the phosphate group on to the amino acid tyrosine in certain key proteins. This **phosphorylation** of proteins is now recognized to have profound effects, modifying the cell's growth pattern and its response to a variety of stimuli coming from the outside. In other words, the cell 'infected' with this oncogene has an altered biochemistry and altered responses to regulatory signals. It has since been found that several other oncogenes are protein kinases. Examples have also been uncovered where a single virus carries two cancer genes into a cell. The first gene causes the cell to divide more extensively in tissue culture, but without yet making the cells so angry that they can cause tumours when injected back into animals. The second gene has no effect when injected into a cell by itself, but, in conjunction with the first one, it completes the malignant transformation. This, too, is important, because we know from clinical studies that human cancer is a multi-stage process, with cells becoming more disregulated in progressive steps. The principles are illustrated in Figure 11.

The final discovery, however, is the most amazing of all. The viral oncogenes, or something extremely similar to them, are present in perfectly normal cells! Moreover, they are not only in the cells of the laboratory animals in which the cancer viruses live.

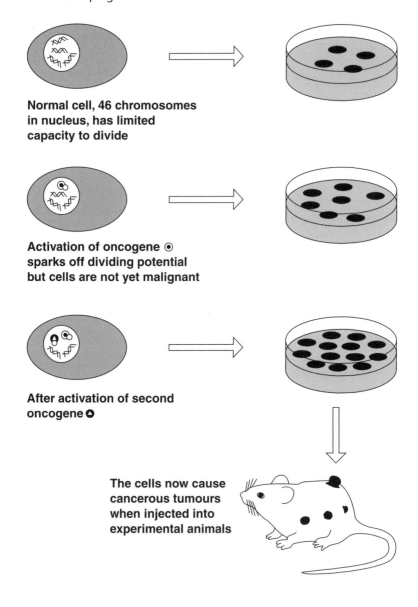

Normal cell, 46 chromosomes
in nucleus, has limited
capacity to divide

Activation of oncogene ⊙
sparks off dividing potential
but cells are not yet malignant

After activation of second
oncogene ○

The cells now cause
cancerous tumours
when injected into
experimental animals

Figure 11. Two possible steps in the development of malignancy
involving the activation of two separate oncogenes.

The activation of the first oncogene confers immortality on the cell, but
does not make it fully cancerous. The activation of the second oncogene
completes the malignant transformation.

Rather, they are widespread throughout the animal kingdom, including humans! Most of the approximately fifty known oncogenes have been faithfully preserved through hundreds of millions of years of evolution, their structures being only subtly different in the different vertebrates. Clearly, these genes must serve some crucial role in normal growth and development, otherwise evolution would not have conserved them so carefully. However, in the normal cell, the level of activity of the oncogene is very low, and so its protein product is made only in minute amounts or not at all. In the virus-transformed cells, the gene is inserted in an inappropriate place and is switched on very actively. Lots of oncogene products are made, and the cell is thrown out of control. Obviously, aeons ago in evolution, a virus picked up an oncogene from the normal DNA of the cell, and began creating mayhem with it as it carried its now lethal passenger to a wrong destination.

The finding that normal cells possess oncogenes, which are called cellular oncogenes to differentiate them from the slightly different viral oncogenes, immediately raises the question of whether such cellular oncogenes can be activated to abnormally high degrees, and whether this might lie behind cancer as we see it in humans. Therefore, genetic engineering has been used to determine to what level oncogenes are activated in various human cancers. A large number of examples have now been found where an oncogene has indeed been grossly activated. At the Walter and Eliza Hall Institute of Medical Research, Doctors Suzanne Cory and Jerry Adams discovered a fascinating example of this. In malignancies of a form of white blood cell called B-lymphocytes, they discovered that the DNA double helix breaks at two specific places. Each of two chromosomes breaks into two. When these breaks are healed, the fragments join up inappropriately. In the process, a particular cellular oncogene called c-myc finds itself not on the chromosome where it should be, but right in the middle of genes, on another chromosome, the proper role of which is to direct antibody formation. There, the c-myc gene is activated to a high degree. Following on these observations, Drs Cory and Adams introduced an activated c-myc gene into mice to create a new breed of **transgenic** mice that expressed the c-myc gene product at very high levels for all their lives. These mice have an extraordinarily high incidence of cancer of the white blood cells, with all mice eventually developing cancers. Because different mice develop the tumour at different ages, it does suggest there are additional factors involved in triggering the cancer. However these mice compellingly prove that activation of c-myc must be critically involved in the malignant process. Breakage and faulty repair of DNA is clearly only one of the ways in which cellular oncogenes can be inappropriately activated. However it is potentially a very important way for the human cancers, because it is known that most of the chemicals which can cause cancer in people accelerate or promote genetic accidents of this sort. Another cause of inappropriate oncogene activation is having extra copies of an oncogene all producing the oncogene product at a low level, but collectively giving a large amount of the protein. In a subgroup of human breast cancer, such a situation occurs. The cellular analogue of the oncogene called neu is present in many copies. The more copies present in the cell, the more aggressively the cancer grows. Tissue biopsy and Southern blot analysis can give a very accurate indicator of the rate with which the tumour will spread.

Cellular oncogenes are present in every normal cell. They do not cause cancer unless they are tossed into the wrong part of the nucleus or are switched on incorrectly. Obviously, nature would not have bothered to preserve such potentially dangerous genes if they did not have an important function in health. The proteins made when some oncogenes are switched on are growth-promoting hormones, rather like those described when Dr Metcalf's work on colony stimulating factors was discussed. Obviously, making too much of such a factor could overstimulate cells and predispose to cancer. Other oncogenes make antenna-like molecules which sense and respond to growth factors. If a cell possesses too many of these antennae, it will grow too strongly even if the amount of growth promoting hormone is quite normal. In some other cases, the oncogenes encode proteins that prevent cell death. So the story of oncogenes is gradually becoming clear, teaching us much not just about cancer but also about normal growth control.

Tailoring of retroviruses for genetic engineering

Retroviruses have two qualities that make them potentially exciting vectors for genetic engineering in mammals, eventually including humans. First, they are usually designed to enter and live in particular types of cells; that is, potentially they could serve as agents bringing foreign genes not to every cell in the body, but only to certain tissues affected by a genetic disease. Secondly, their reverse transcriptase and the associated control machinery could allow a simple linear polymer of RNA to become integrated into the DNA of the cell. Therefore, scientists are currently modifying retroviruses in interesting ways. They are cutting out the dangerous part, the retrovirus oncogene, and putting in instead desired genetic information, in RNA form. Successes have already been achieved in genetic engineering of tissue-cultured human cells by this method. Other viral vectors that are being studied include the AIDS virus and some common viruses such as adenovirus, normally a cause of respiratory infections and herpesvirus, normally a cause of nerve infections. Again it was essential for scientists to remove all parts of the virus that can do harm. Once this was done, a human gene was inserted in the vector using the techniques we described in Chapter 3. The choice of gene will of course depend on the disease to be treated. These **gene therapy** vectors are now undergoing various types of clinical trial for safety and efficacy in patients.

DNA can also be inserted into mammalian cells in other ways. A simple method involves packaging the DNA inside a bubble of fat. The fat fuses with the **cell membrane** and the relevant DNA including the gene to be introduced enters the cell. Alternatively, the DNA can be injected into cells by coating gold pellets with them and firing the pellets into the cells. Finally instead of inserting a new gene, scientists are investigating methods of repairing the mutations in the incorrect genes. The reasoning goes that cells already possess a method for proof-reading the sequences of genes to ensure they do not mutate. If we could utilize these processes, we would be left with a normal gene, but importantly it would be sitting in the right place in the DNA and be under the correct sort of control. So scientists have been investigating putting in gene fragments with the correct sequence and trying to get the cell's mismatch repair systems to make the changes. Called 'chimeroplasty', this technique is showing a surprising ability to do the job in early tissue culture experiments. Thus, in contrast to the late 1970s when genetic engineering involved only bacteria and viruses, we now have an ever-growing series of techniques for getting foreign genetic information into animal cells. Some of these ways integrate the new gene into the cell's own DNA, though we cannot yet control just where the transferred gene is placed, except with the method of chimeroplasty. Other methods leave the transferred gene functioning and replicating independently in the cytoplasm of the cell. The methods currently available, although much improved, are not yet perfect—they should be regarded as evolving.

So we have a method of introducing genes, but how would we actually use them? Well, there are two broad approaches. One is to remove cells from the body and to expose them to the gene therapy treatment. We would grow these cells in culture for a while to check that the gene had gone in and then return them to the patient. The second method would be to inject the genes directly into the patient. There we would have less control over where the genes end up. Attempts are being made to target the genes to specific cells, where we would want the new protein to be made. This is done using the virus vectors by coating the virus particles with the special proteins that viruses themselves use to target certain cells. Thus if we wanted the cystic fibrosis genes to go to the cells of the lung, we might coat the gene therapy viruses with proteins from viruses that cause bronchitis. These viruses attack cells that line the air passages.

So this is the theory. Let us see how it all works in practice. A great deal of experimentation over the last fifteen years has shown us the major difficulties that need to be overcome before gene therapy can be used in the clinics. There are at least three major problems that are largely responsible for the disappointing results to date. The first of these is that although most of our methods do a pretty good job of getting DNA into dividing cells, they are not so good at getting DNA into resting cells. Now by resting, we do not mean that the cells are doing nothing. They are engaged in the ordinary day-to-day, metabolic processes, but they are resting with respect to cell division. Technically they are in a non-dividing part of their cycle called G_0, where the enzymes that copy DNA are largely turned off. Most cells in the body are in G_0. If our introduced gene does not get incorporated into the cell, then it will not be very effective in making new proteins. Secondly, when we introduce a virus with a gene that makes a new protein, we are exposing the body to a sort of artificial infection. The immune system cannot distinguish this as something potentially beneficial to the body, but regards it as a viral infection. It therefore sets out to eradicate the cells that have been infected with the gene therapy virus, just as it eradicates virus-infected cells when we get a cold or the 'flu. Remember also that in some cases we are using gene therapy to provide a protein that the patient does not make because of an inherited disease. That person's immune system will not realize that the newly introduced protein is of human origin. Again it will see it as something foreign and it will be the target of a ferocious immune attack. Finally, cells have the capacity, again probably developed as part of defences against viruses, to turn off genes within the cell that are being expressed at high levels. Thus we see experimentally that after a while the introduced gene falls silent, and new protein production ceases.

There are strategies to try and overcome all of these problems. For example, the AIDS virus is capable of infecting cells in G_0, so an appropriately engineered virus of this sort could infect resting cells. The use of 'naked DNA' would mean that the body would not make any immune response to viral infections, and so on. We are seeing now a resurgence in the number and scope of clinical trials, but still the final goal remains elusive. This is very much a developing therapy, but given the paucity of alternatives in the case of many of these diseases, its further development is fully justified.

The most successful demonstration of the use of gene therapy comes from a study on a rare inherited disease which leaves the unfortunate victim without a functioning immune system. The disease, called severe combined immunodeficiency-X1 disease, occurs in people who possess a defective gene for a protein required for the development of T lymphocytes (also called T cells). The defective gene was placed in a virus and used to infect patient cells in a test tube. The cells were returned to the body and started to develop normally. Some ten months after transfer, the two patients who were children eight and twelve months old at the time of therapy had developed normal numbers of lymphocytes and appeared to have functioning immune systems capable of protecting them from disease. There are some special features about this disease that favour the success of the gene therapy and apart from this study, there is little published proof of convincing long-term effects in people or an improvement in clinical state. Moreover there are dangers with this form of therapy, as with almost all medical interventions. Just recently we have seen a tragic death occurring during the treatment of an inherited liver disease, when the viral vector being used caused tissue damage when infused into the liver. The patient died of liver failure. In the subsequent investigation it appeared that the researchers involved had failed to abide completely by safety guidelines. This tragedy emphasizes how we are still in a very early stage of development of gene therapy and, although we favour continued development, it is essential that a careful and cautious approach be adopted.

How will gene therapy be deployed?

There is no shortage of problems to be addressed. Scientists' attention will be directed first at the diseases which are clearly genetic in nature—diseases like the haemoglobinopathies, **haemophilia,** nervous system conditions such as phenylketonuria and **Tay-Sachs disease,** and a host of other biochemical disorders. They will also certainly home in on the challenges flowing from the discovery of oncogenes. Ways of turning off the oncogene or of neutralizing its product will be sought eagerly as a new approach to cancer therapy. Though it presently strains credulity, one can even dream of vaccinating people against the products of oncogenes and thereby preventing cancer—a dangerous thought as oncogenes clearly have normal functions, but conceivably some of these functions are over and done with in adult life. Scientists will also learn more about the genetic component in diseases like diabetes and multiple sclerosis, though it is presently not easy to see how these could be addressed in a manipulative sense. Other approaches will seek to infuse genes for protein products that are no longer being made by the body because some specialized tissue has been destroyed by disease. Taking diabetes as an example again, it would not be outrageous to conceive of inserting insulin genes and switching them on inside some cell other than a pancreatic cell. The problem then would be to design a requisite control system so that the correct amounts of insulin are made at all times, a formidably daunting task. These examples, somewhat akin to science fiction, are mentioned only to indicate that the implications of genetic engineering for medical therapy are indeed very open-ended.

If one were to look fifty to a hundred years ahead, to a period where many genetic diseases were being cured in children and adults, it would be legitimate to voice a concern about the implications of this for the gene pool of the human species. At present, the worst of these diseases do kill before people reach child-bearing age. If the disease were cured through engineering of a particular tissue, say the bone marrow, the person concerned survives to have children, but as the ova and sperms have not been treated, the bad gene is passed on to the next generation. Donning science-fiction glasses once again, there would be an approach to that dilemma. The fertilized ovum is likely to be a good target for genetic engineering. The patient, then, would have to be persuaded to have children only through **in vitro fertilization**. The genetically cured fertilized ovum would then grow into a normal individual, whose sperms or ova, in turn, would be genetically normal, thus breaking the chain of poor genes handed down to future generations. Presumably a society sophisticated and affluent enough to cure the genetic diseases in the first place would also be in a position to take this further step.

For some of the genetic diseases we have considered, there is a clear association between abnormality in a particular gene and the disease. What about those cases where the association is weaker? There are now a number of genes associated with increased risk of heart disease, or particular autoimmune diseases such as diabetes or rheumatoid arthritis, or neurological diseases such as Parkinson's disease or Alzheimer's disease. Would it make sense to alter these genes to variants associated with protection against these diseases rather than susceptibility? We would strongly make the point that such ideas are extremely premature. A form of a gene that may be associated with risk for a particular disease may actually be very beneficial under other circumstances. One example from Nature may be worth citing. Malaria in humans is caused by a number of species of the protozoan parasite *Plasmodium*. One of these, *Plasmodium vivax,* invades red blood cells by attaching to a blood group protein, called the Duffy antigen, on the surface of the red blood cell. In parts of Africa, most people lack the Duffy antigen and so are genetically immune to *P. vivax* malaria. They can never catch this form of the disease. It might be argued that perhaps this is a way to lessen the burden of malaria by using gene therapy to inactivate this disease. In fact, the Duffy antigen is also involved in the immune response, acting as a receptor for particular hormones of the immune system called chemokines. It now appears that these same Africans have more severe bacterial infections as the price to pay for avoiding malaria. It is likely that similar sorts of checks and balances operate in the case of variable genes. They may confer a risk in some situations and a benefit in others. A very deep understanding of their roles will be required before we start to re-engineer people.

Moving beyond the case of single genes, it is important to point out that genetic engineering does not seek to address complex multi-gene traits in complex multicellular organisms. The possibility that some mad dictator might use the technology to create a superman of great strength or intelligence (or whatever) is out of the question. We know precious little about the genetic basis of the human qualities we value. It is vital not to let our understanding of the highly precise things that can be achieved be clouded by fanciful notions and fears. The use of genetic engineering in creating microbes of greater virulence for purposes of biological warfare is a much more sensible thing to worry about, because given sufficient effort, this might be achievable. As much surer weapons of mass destruction already exist for the superpowers, biological warfare research is probably confined to a few rogue states, and unengineered pathogens such as smallpox virus and anthrax bacteria appear to be the favoured weapons. Nevertheless, it is important to take mankind's nightmares seriously, and we shall return to these issues.

A final matter which should be mentioned is that of cost. At the moment, the costs involved in some of the most elaborate things which medical scientists do for patients are staggering. A heart transplant with all the pre- and post-operative care required costs around $100 000. Initially, the costs of whole body irradiation and transplantation of genetically cured bone marrow would probably be greater still. Are these costs so immense that we would be better off making the decision right here and now not to pursue these lines of research? The answer to that question is an unequivocal no. At the beginning, we grope for halfway house measures that are frighteningly complex, because we do not know enough. As we learn more, we will see that these early attempts represent only imperfect, compromise solutions, and we will find bolder and more elegant approaches which will also be cheaper. For example, if we do succeed in fashioning a virus which homes in on red blood cell precursors, curing their haemoglobin genes, and if we omit the amortization of the research costs, we might have a cure that costs no more than a simple vaccine shot! In the long term, the cost issue should not be a barrier to gene therapy. In the short term, there will be some very tough decisions about the selection of patients on whom doctors learn how to turn the bench scientists' dreams into hard clinical realities.

Stem cell therapy—cellular replacements for malfunctioning tissues

The potentialities of gene therapy are exciting indeed, but there are many human illnesses that are due to more than just a faulty gene. Even if we were willing and able to extend gene therapy to the area of disease susceptibility genes (see Chapters 4 and 6) there are conditions for which gene therapy would not be useful. For example, there are a number of genetic factors that predispose to heart disease. We may contemplate changing the genes of a person who is prone to heart attacks because of their genetic propensity to have high levels of cholesterol and other fats in the blood. If the healthier form of the gene could be successfully introduced, then this would lower their risk of heart disease and over time they would be less likely to build up fat deposits (atheroma) in their blood vessels. But suppose they had a heart attack and severely damaged the heart muscle. No amount of gene therapy adding one or a couple of genes would help this person recover. The muscle cells in this person's heart have died and been replaced by fibrous tissue. We have no idea how to add genes to these fibrous tissue cells and turn them back into muscle. It is likely that hundreds to thousands of genes are acting differently and would have to be switched off and on appropriately. What if the heart muscle was so severely damaged that drugs were no longer useful and the only help would be a new heart? We could supply this only by the difficult and costly process of heart transplantation. But what if instead we could treat the patient in a way that would grow a new heart in place of the old? Incredible as it may seem such a prospect is now visible on the distant horizon. Called **stem cell** therapy, it is being applied in animals with exciting preliminary results.

Stem cells are cells that replenish themselves while also giving rise to cells with different functions. There are different sorts of stem cells with different potentialities. To understand them we will start from the fact that our bodies originally grow from a single cell, the fused egg and sperm. This fertilized egg has the potential to develop all the different tissues, whether it be heart, liver, brain or bone, needed to form a complete person. The term used is **totipotent**, essentially meaning it has total power to make all these tissues. As the fertilized egg divides, at first all the daughter cells retain this unlimited capacity to make new tissues and organs. If the ball of totipotent stem cells should split in two, then each can give rise to a complete person, and in fact this is how identical twins occur. However after about four days of division, the cells start to become more specialized. Daughter cells in the outer layer start to give rise to the placenta while cells in the inner layer give rise to the developing foetus. These inner layer cells will go on to form essentially all of the tissues of the human body. They cannot however form a whole human being any longer, because they have lost the ability to form a placenta, and the other tissues needed to support development in the uterus. These cells are called **pluripotent** because they retain the capability to form very many cell types. However it should be stressed that their potential is not total; they are not 'totipotent' and they are not embryos, nor can they form embryos.

After some further cell division, the pluripotent cells begin to specialize further and now give rise to cells that have a particular function. They might become blood stem cells that can give rise to all the cell types found in blood. These include red blood cells, all types of white blood cells and platelets. Other specialized stem cells might give rise to various sorts of cells found in the brain. These more specialized stem cells are called 'multipotent'. We can start to see now how if we had a multipotent stem cell that made heart muscle, then infusing this into a person might be a way to rebuild a damaged heart. Outside a laboratory, this process of specialization from totipotent stem cells, to pluripotent stem cells to multipotent stem cells goes in this direction only. Pluripotent stem cells cannot produce totipotent stem cells and so on. Stem cells don't only occur in the developing foetus. Stem cells in the adult bone marrow are constantly providing us with new blood cells as we wear out our red and white blood cells.

So now that we understand the different types of stem cells, what uses could we put them to? Initially, possessing such cells would help us do experiments that explain how humans and other animals develop. We could start to learn how cells make decisions to specialize and which genes are required for this to happen. Since major health problems such as cancer and birth defects are due to abnormalities in this process, if we understood it better, we might be able to develop new therapies to combat them. Secondly, the possession of human stem cells would allow us to test new drugs for safety and efficacy in a variety of tissues before animal and human testing. It is likely that such tests would give us a better idea of which drugs are worth proceeding with. This would make drug development cheaper and could also decrease the use of animals in drug testing.

As already alluded to, such stem cells might allow us to generate tissues for introduction into patients. Possibilities include all of the conditions that are currently treated by transplantation, whether it be to replace kidneys, livers, hearts, lungs or corneas. Whenever an illness has resulted in damage or destruction of a tissue, we would have an opportunity to use stem cell therapy. Other conditions that might benefit include nervous system diseases such as stroke, Alzheimer's disease, Parkinson's disease and spinal cord injury due to trauma such as car accidents. In diseases such as osteoarthritis and rheumatoid arthritis, the cartilage in the joints is destroyed. Cartilage stem cells could potentially grow back this tissue and restore joint function and remove the terrible pain associated with these conditions in their severe forms.

Are these ideas anything other than a pipedream? In the case of damaged hearts, it has already been shown in animals that heart muscle cells transplanted into a damaged heart can start to re-populate the heart tissue, and working together with the remaining host heart cells, improve cardiac function. In other experiments, researchers at Washington University in St Louis have shown that when they injected immature nerve cells derived from mouse embryonic stem cells into rats whose rear legs had been paralysed by damage to their spinal cords, the animals regained some mobility. One month after receiving the cells, the rats could lift their rear ends and take steps with their rear legs, even if somewhat awkwardly. Rats who did not receive any cells could not move their rear legs at all. The implications for the treatment of spinal cord injury are astonishing.

What are the impediments to starting such therapy now? Clearly, we need a source of human stem cells that can be used for experiments and therapy. In fact there have already been two successful attempts to collect stem cells, but also a great deal of attendant controversy. In both cases, pluripotent stem cells were collected from the inner cell layer of human embryos. In one case this was from embryos in an IVF clinic that were in excess of those needed for fertility treatment. In the second case the stem cells came from the gonadal region (that region of a foetus destined to become the sex organs, either the ovaries or the testes) of foetal tissues resulting from therapeutic abortions. In both cases appropriate informed consent was obtained from the people involved and ethical guidelines were followed. However it should be noted that for many people, there is no ethical way in which such tissues could be obtained and the termination of a foetus by a therapeutic abortion or destruction of IVF embryos is tantamount to murder. We do not share such a belief but recognize that this issue must be adequately debated before the full-scale study of this area. It might be that obtaining such cell lines may only need to be done very infrequently, as they should have the potential to self-replenish. Nevertheless, it is an area that causes great emotional anguish in many and we must proceed sensitively.

Secondly, we need to understand more of the process by which cells specialize. It would be terrible to infuse stem cells, if they continued to grow in an uncontrolled manner, giving rise to a cancer in the patient. What if we infuse cells hoping that they will repair the heart and instead they turn into liver cells? The treatment would be useless and perhaps dangerous. We must know more about what controls this process. Fortunately, extensive work in mice and other experimental animals is rapidly extending our understanding here.

Thirdly, these cells face the same potential problems as organ transplants, that of rejection by the immune system. As cells they will have the same tissue typing antigens as do organs and must be perfectly matched to the recipient in order not to be destroyed. We would need to create banks of stem cells of all the different tissue-type profiles for use, a formidably difficult and expensive task; or perhaps work out ways to mask the tissue type antigens so as not to provoke attack. Alternatively, perhaps we can develop a method to make stem cells from the patients themselves. In the next section, when we discuss human cloning, we will discuss the technique of **somatic cell nuclear transfer**, a technique used to clone animals. Applied to humans, it could produce a source of stem cells, again only if ethical issues can be resolved.

So somewhat like gene therapy, stem cell therapy would appear to have extraordinary promise, but many problems remain to be solved. We would appear to be even further away from the use of this therapy in the clinic than gene therapy. However, the first fledgling steps are being taken.

Human cloning

The beginning of the new millennium has been attended by a wholesale review of what we have achieved in the past and a host of predictions of what we will achieve in the future. Much interest has focused on the issue of cloning of animals and humans. In a later chapter we will discuss the construction of genetically altered plants and animals, but for now let us focus on the cloning of humans. We have already come across the term cloning when we discussed the basic techniques of genetic engineering. By cloning we mean the production of identical copies of an organism by a non-sexual process. As we have already pointed out, the process of fertilization is preceded by a shuffling of the genetic material of father and mother, so that any offspring is a mosaic of genes inherited from both parents. A clone in contrast would have an identical gene set to the parent organism and would be as similar to that parent as identical twins are to each other.

How is this process possible? In plants it is often accomplished by taking a cutting and growing up a new plant from that cutting. In animals it is more complicated because, as we discussed in the previous section, animal cells lose the capacity to produce all the tissues of the body in their correct shape and location. Thus if we were to take a piece of an animal or a human and try and culture it, the outcomes would be that either the cells would die or they might go on growing for a while but produce only a limited number of new cells. A piece of skin might produce further skin cells, but not cells of the pancreas, liver cells might produce more liver cells but not nerve cells. This is because, as already noted, cells in the body go through a process of differentiation during development in which certain genes are turned off and the cells often lose the capacity to divide. There are those rare cells we mentioned called stem cells with a greater potential for growth but, unless they are totipotent stem cells, they are already adapted to make a more limited number of cell types. Nevertheless, all cells do contain a full complement of genes, and if these can be made to function in the correctly co-ordinated fashion, then in theory a new organism could be made. This new organism would have the same set of genes and would be a clone. Theory says that cloning should be possible and it was left to the ingenuity of scientists to determine the means by which this could be done.

It turns out that different organisms are more or less resistant to cloning. For bacteria, cloning is the normal way of reproduction. Amphibia have extraordinary regenerative capacity and over 30 years ago Dr John Gurdon managed to clone frogs. It is possible to take very early foetuses and split them to make more identical copies, in a process similar to the way identical twins form. However the feat of taking an adult mammal and making a clone proved to be surprisingly difficult, until the breakthrough made in 1996 by Dr Ian Wilmut at the Roslin Institute in the UK. He and his colleagues used a technique in which they removed the nucleus from a fertilized egg and continued to grow it in tissue culture. Left to itself this damaged cell would do very little, as it lacked the necessary genes to instruct the making of proteins. These scientists fused this egg cell with an adult cell taken from the udder of a sheep. Now cell fusion of this sort had been attempted before with adult cells, but the results were unimpressive. However, by careful choice of the experimental conditions Dr Wilmut managed, by starving the adult cells, to convert them back to a state similar to a totipotent stem cell. In some instances, the fused cells started to divide, formed an embryo and were implanted into the uterus of sheep chemically treated to be in a 'pregnant state'. One of these implanted embryos developed normally and sheep 6LL3, otherwise known as Dolly, was born. Now the process was very inefficient with only 1 of 277 attempts actually working. However, once cloning of adult mammals had been demonstrated in principle, the method could be refined, made more efficient and applied to species other than sheep.

Where do we stand at present? Research on Dolly has shown that in fact she did not start life as a normal lamb, but had a cellular age of approximately six years old, the age of the sheep that donated the udder cells. How can scientists establish this? It turns out that chromosomes, those large pieces of DNA found in our cells, have very specialized structures at their ends called telomeres. These are specific short sequences of DNA repeated many times that stop the DNA at the end from being digested inappropriately. As cells divide these telomeres become shorter and it appears that when they get very short, the cells can no longer divide. One of the reasons cancer cells continue to divide is that they make a special enzyme that lengthens telomeres. When the chromosomes in Dolly's cells were examined, the telomeres appeared shorter than would have been expected from her chronological age. This suggests that it would not be possible to clone an individual, then when that individual ages, clone again and so on to achieve immortality. Presumably, the accumulated telomere shortening would eventually lead to cells that could no longer reproduce. Another problem for clones is that as we live, damage accumulates in our DNA from chemicals and exposure to cosmic rays. This damage is often removed from the genes during fertilization as DNA is swapped between egg and sperm. In a clone, this damage would accumulate and never be repaired. Experience is starting to accumulate that suggests that clones are prone to a number of defects and the production of healthy animals is proving difficult.

Is cloning a technique that only works in sheep? Since the cloning of Dolly, we have had the announcement of the cloning of cows, pigs, goats and mice using similar techniques. Some of the cloned animals were produced from a transgenic parent, i.e. an animal that had an introduced gene. We discuss in detail the science behind the production of transgenic animals in Chapter 9, but for now think of it as an animal with an extra gene introduced into it. The gene may come from any other living organism. Researchers at Genzyme Transgenics in Framingham, Massachusetts, have produced a goat that makes the human protein antithrombin III. Antithrombin III is a protein that can be used to prevent damaging blood clots after strokes or heart attacks. This goat was made by injecting fertilized goat eggs with the gene for antithrombin. When the female foetuses were forty days old, researchers fused foetal cells with goat eggs whose DNA had been removed. The eggs were then put into females. The foreign gene causes the protein to be made in the goats' mammary glands and secreted in their milk. The antithrombin, which is hard to make by other means because of its complex structure, can be readily purified from the milk. Note that this process did not involve cloning an adult as had been done with Dolly, but the technically easier feat of cloning foetal cells. A Canadian company called Nexia has produced goats that make a protein normally made by orb-weaving spiders in their dragline silk. The protein has a very high tensile strength and is biodegradable which suits it for a great range of medical and engineering uses, including sutures. Goats are abundant milk producers and don't take as long as cows to mature, making them a better choice for rapid production of these proteins. We will undoubtedly see cloning used in agricultural techniques as it allows the rapid expansion of a single valuable animal into a herd of identical progeny.

The announcement of the successful production of Dolly caused a furore. The technique of cloning and its potential application to humans seemed to speak to our deepest fears about the human condition. A common reaction was that cloning could be used to create a super race, perhaps a race of Adolf Hitlers who could take over the world. Alternatively, the technology could be used to clone Albert Einstein or Mother Teresa or even, as a recent poll suggested, Michelle Pfeiffer. Would any of this be possible or is it totally in the realm of science fiction? If possible, should it ever be attempted or should it be banned forever?

The techniques used to make Dolly and the other animals could in theory be applied to any mammal including humans. There may be special problems attendant on the use of human cells, but there is no reason to think that they would be insurmountable. Thus the only bar to the performance of human cloning is likely to be moral suasion and legislative and administrative bans. In the United States, President Clinton referred the matter to the National Bioethics Advisory Commission, which concluded in June 1997 that at this time it is 'morally unacceptable for anyone in the public or private sector, whether in a research or clinical setting, to attempt to create a child using somatic cell nuclear transfer cloning'. In the United Kingdom, the Human Genetics Advisory Commission and the Human Fertilization and Embryology Authority have recommended that the government should enact a law outlawing reproductive cloning of humans. A number of other countries have announced laws banning human cloning and all the major agencies that support scientific research in industrialized countries have announced bans on the use of their funds for such experiments. The sad truth is, however, that the techniques are not excessively complicated and could be performed by any clinic that is capable of performing in vitro fertilization techniques. The equipment required is also relatively cheap and at present the only drawback appears to be that the process is very inefficient. This unfortunately makes it probable that at some time human cloning will be performed in some place, public opprobrium notwithstanding. Reports have already appeared claiming that human cloning experiments were attempted in South Korea during 1998. It is unclear whether cloning in fact occurred and we have no way of authenticating the reports, but it serves to remind us that the relevant technology might be available in countries where mechanisms of control of medical research have not yet reached the elaborate levels of countries such as the USA, the UK or Australia.

If such a thing were possible, what would a human clone be like? In practice, it would not be absolutely identical to the parent, as environmental factors such as the hormones and blood supply in the uterus affect how a foetus develops. The closest analogy to how similar a clone might be to its parent is to consider how similar identical twins who have been raised separately are to each other. A large number of such twin pairs have now been studied, and although they are strikingly similar in some ways, in others they are different. They are physically similar, but their personality traits will on average have about a 50% correlation, i.e. they are generally similar but certainly not identical. This level of similarity compares to fraternal twins who show a 25% correlation, non-twin siblings who show an 11% correlation and unrelated strangers whose personality traits show low correlation close to zero. Given that clones would not share the same intra-uterine environment as the parent, it is likely their personalities might be somewhat less than 50% similar to the parent. This sort of information makes scenarios of cloning a group of totalitarian warriors less realistic. In fact, given the history of our century, there seems little reason to go to such lengths. We have seen that entire countries have been persuaded to follow the wishes of their leaders, no matter what the consequences. Nazi Germany, Stalinist Russia and Cambodia under Pol Pot are obvious and by no means unique examples. Cloning seems an extraordinarily inefficient way to exert such control, and we do not believe it likely that cloning technology would be used in such a way.

It is more likely to be deployed in individual cases to meet the desires of those in unusual circumstances, perhaps parents whose only child is dying of an incurable disease, but who can no longer have children. Cloning that child would provide the parents with a second chance at parenthood. Or consider a widow whose husband has died tragically before they could have children. It is possible to construct heart-rending scenarios where human cloning would provide a means of assuaging human grief. It raises many questions that can only be resolved by continued debate. We would simply say at this point that in our view, the cloning of humans in the hope of generating a particular kind of person is not acceptable. Children should be born into a situation where they have freedom to develop and not be pre-ordained to be a mathematical genius or a champion basketball player. Secondly, there must be some limit to the degree in which medicine and technology should be used to further reproduction of the human species. We do not suffer from a shortage of people in this world, and we believe our resources should be addressed to ensuring a decent quality of life for the children who already live on this planet. Biotechnology can aid in that desire, but perhaps not by perfecting the technique of cloning.

8

Vaccines of the future

Vaccines are the world's most cost-effective public health tools. Today, we take childhood immunization programmes very much for granted, scarcely remembering that epidemics and plagues were once the stuff of history: key determinants of the fate of individuals, the winning or losing of wars and the success or otherwise of large-scale migration. Our memories were jogged somewhat in 1980 when the World Health Organization officially declared the success of its smallpox eradication campaign—this dreaded scourge having been conquered completely by global immunization. For the first time ever, a disease had been totally and permanently eliminated. But in a world accustomed to change, yesterday's dramatic achievements become today's commonplace with frightening speed. Many would believe that infections have largely disappeared as an important cause of disease, perhaps with the exception of AIDS, which is now well controlled by drugs. In fact it may come as a surprise to learn that infectious diseases are responsible for almost half of all disease and death in the world. Scientists have continued to study infections in the hope of supplying a solution as effective as the smallpox vaccine. These studies are now approaching fruition and we are on the threshold of a leap forward in the field of vaccines at least as important as the last major quantum jump when poliomyelitis vaccine was introduced in the 1950s. The vaccines of the future will be the products of modern biotechnology.

Principles of successful immunization

The immune system is nature's way of defending vertebrate species against infectious diseases. Tragic examples of what happens when the immune system fails are seen in various illnesses, congenital immune deficiency or the acquired immune deficiency syndrome (AIDS), for example. The end result is death from overwhelming infection. In normal individuals the immune system is provoked to form *antibodies* when foreign organisms enter the body and multiply within it. Sometimes when a virulent microbe infects humans, the antibodies are formed too late, and the patient dies. On other occasions, the organism concerned has, through evolution, devised clever tricks of evading the host's defences, resulting in a chronic disease like tuberculosis or schistosomiasis, despite the formation of antibodies. Very frequently, however, the antibodies both vanquish the first infection, and leave the patient immune against that particular disease for long periods or even for life.

There are also some infections where antibodies are inadequate to control the disease but where a different kind of immune response does the job. Antibodies are protein molecules which travel in the bloodstream, and they are specific for the infecting microbe—measles antibodies prevent measles, not poliomyelitis. However antibodies must find their target out in the open and if microbes hide within cells, the antibodies are powerless to help. The second tool of immune defence is the T cell, and these have been developed to look inside cells and find the microbes that have taken haven there. T cells are born in the thymus, and exert their defence function in a strictly localized manner. For example, measles-specific *cytotoxic* T cells kill measles-infected cells, thereby impeding spread of virus, whereas measles-specific *helper* T cells secrete powerful inflammatory molecules when they encounter the measles-infected cell, again helping the body to limit spread of the disease. Immunity based on T cells is called cellular or cell-mediated immunity.

The key principle which unites all forms of successful immunization is devising a way in which the formation of specific antibodies (and/or cellular immunity) can be provoked without the person or animal concerned having to run the gamut of an actual infection. In the best cases, this leaves an immunity just as good as that found in a person who has had the disease and has recovered from it. With some other vaccines, the protection is less perfect but still substantial, so the risk of getting the relevant infection is much reduced and the disease itself is less severe in those people who do come down with it. Vaccines work because the cells (white blood cells called lymphocytes) which make antibodies or which mediate cellular immunity need not interact with living, virulent micro-organisms in order to be provoked into action. The lymphocytes' capacity to form the protective immune response is triggered when they encounter specific molecules coming from the foreign invader. These molecules are known as *antigens*. So all immunization involves introducing antigens into the body in a risk-free manner. The antigens soon reach the lymph glands and the spleen and there trigger division among lymphocyte cells, which start to form antibodies within a few days or else produce an army of angry T cells.

Broadly speaking, immunization has been accomplished in one of three ways. The first, which the Gloucestershire general prac-titioner Edward Jenner stumbled across in 1796, was developed much further by Louis Pasteur in the late nineteenth century. The method is to find a harmless relative of a virulent organism some-where in nature, or intentionally to change a virulent microbe into a harmless strain through prolonged culture outside the body or repeated passage through different host animals. These procedures may throw up a mutant, non-virulent organism which can then be allowed to grow and multiply within the body, and thereby provoke antibody formation. Such live, attenuated vaccines work because the harmless relative and the virulent organism share important antigens, and therefore the antibodies against the relative can also attack the real organism when it comes along. The second method involves killing the virulent organism, for example with formalin, and injecting it into the body. The antigen molecules from these killed organisms can be effective at surprisingly low doses, as the brilliant success of the Salk poliomyelitis vaccine proved. However, killed vaccines are usually given as two or more injections to ensure that the stimulus to the immune system is sufficiently strong. The third method rests on the fact that one does not have to be immune to every antigen of a microbe in order to be protected. One can therefore inject some *component* of the micro-organism rather than the entire living or killed microbe. The current highly successful diphtheria and tetanus vaccines work on this principle. In those cases, the operative antigen is a modified version of a toxin that the relevant bacteria may produce, but in other cases the antigen might, for example, be a molecule sitting on the outer wall of the bacterium or virus. The purer such molecular vaccines are, the less likely they are to have irritating or dangerous side effects.

Principles of vaccine manufacture

Until recently all vaccine manufacture has involved the large-scale growth of the responsible organism, or its harmless relative, under controlled laboratory conditions. For bacterial vaccines, like those against whooping cough, tetanus or diphtheria, this is relatively straightforward as bacteria can grow in nutrient broths rather like a rich meat soup. For viral vaccines the technology is more demanding. Viruses are smaller forms of life which can grow only inside a living cell. So the vaccine has to be prepared either in living animals (the skin of calves for the smallpox vaccine, or the inner linings of a chick embryo for the yellow fever vaccine) or, more usually, in mammalian or chicken cells that are themselves growing under artificial conditions through the technique of tissue culture.

A great deal of technology has to go into conventional vaccine manufacture. Obviously the growth medium must not become contaminated with even one irrelevant micro-organism, so superb aseptic techniques must be used. The workforce must be carefully protected from dangerous bacteria or viruses. For killed vaccines, every last microbe must be killed. For live, attenuated vaccines, the organisms must be kept in a medium which ensures their continued survival, and frequently this requires cold storage. For molecular vaccines, the right antigen must be purified from all the irrelevant material. Quality control procedures must be stringent and each vaccine batch must be tested for safety and efficacy. All this adds to costs.

The potential of biotechnology in vaccine manufacture

Modern biotechnological advances have unblocked the central bottleneck in vaccine manufacture, namely the need to grow vast quantities of pure and possibly virulent organisms. Frequently, as in the case of tuberculosis or leprosy, the organism is very fastidious in its growth requirements. Some disease provokers, for example

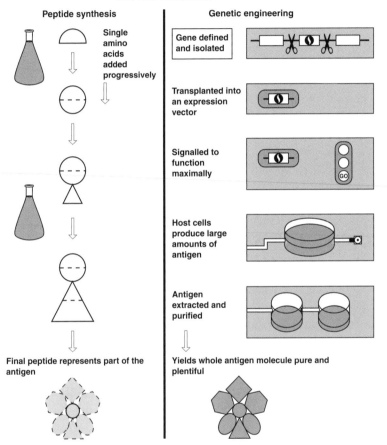

Figure 12. Two separate methods for vaccine manufacture: peptide synthesis and genetic engineering.

Peptide synthesis involves chemical coupling of one amino acid after another to build up an antigen, which frequently is only a small part of a natural protein. Genetic engineering involves transplanting the gene for the antigen into a suitable host cell, as described in Chapters 3 and 4.

the malaria parasite, require large quantities of human blood for their growth. This makes mass production impracticable. Two separate methods can beat these constraints, and indeed are seen by many as competing with each other. One method is to synthesize antigens chemically. The other is to force harmless, easy-to-grow bacteria or yeasts to make antigens through genetic engineering (Figure 12).

The synthetic approach makes antigens in the test tube from simple chemical building blocks. Many antigens are proteins, which, as we saw in Chapter 2, are strings of smaller molecules, the amino acids, hooked together in a particular sequence. The cell does this very quickly and efficiently. The synthetic organic chemist can also, rather more laboriously, build up a protein, step by step, by piecing together the right amino acids. In many cases it is not necessary to inject a whole, intact antigen molecule in order to induce a good immune response. One little corner of a protein, say ten amino acids in length, may suffice to give protection, though, as we shall see, some tricks have to be used to make this work. Whole protein molecules can also be made synthetically from amino acids, but the bigger the protein, the greater the risk of introducing an error into the sequence and the more cumbersome the synthesis. These difficulties mean that, in practice, synthetic proteins are rarely larger than fifty amino acids. Therefore, much emphasis is going into defining 'immuno-dominant' portions of large antigens of medical importance, smaller bits of protein (called peptides), usually eight to twenty amino acids long. Peptides are what the T cell recognizes, and are thus important in vaccines which seek to introduce cell-mediated immunity.

Genetic engineering, the alternative strategy, harnesses living organisms to mass-produce antigens vicariously, exactly as outlined in Chapter 5. The transplanted gene can be big or small, so that the protein made can be of almost any desired size. Of course, it is still necessary to purify the protein made by genetic engineering from all the other molecules inside the *E. coli* or whatever other host organism is used. Recently, another dramatic breakthrough in biotechnology has provided an elegant solution to this problem. Antibodies, as well as being protective substances, are also chemical entities of exquisite specificity. Nature has patterned the surface of the antibody molecule to perform recognition tasks in a highly discriminatory manner, so that two molecules which look very similar to each other can still be picked apart by an antibody raised against one of them. The most refined kinds of antibodies are now made in the test tube by taking one single antibody-forming cell from a mouse and artificially fusing it to a cancer cell. The hybrid cell is now immortal and can grow forever, provided adequate nutrients are present. It continues to make the antibody, called a **monoclonal antibody** to describe its origin from one cell. Once a scientist has obtained a monoclonal antibody to a protein, a giant stride towards cheap purification has been taken. The antibody is bonded to a solid support, for example tiny beads packed into a long thin glass tube, constituting a vertical column. The impure solution containing the protein and a whole host of other molecules from *E. coli* is poured on to the column. The antibody on the beads binds the protein of interest. The scientist then flushes the column through with a large quantity of fluid, removing all the molecules bar the one on the beads. Then, an acid solution is added, and this undoes the bond between antibody and antigen. The wanted protein drips out as a pure solution.

Advantages and disadvantages of the two most commonly used techniques for vaccine development and production

Both the peptide and the recombinant DNA approaches have their ardent proponents. What is not revealed in many such discussions is that the two approaches are very interactive: genetic engineering may be the way to find the small piece of protein that you eventually wish to synthesize, and testing immune responses against small synthetic antigens may help you to validate the importance of a particular large antigen as a candidate vaccine molecule, which you then make through genetic engineering. In practice, many laboratories probing for new vaccines use both techniques in their research. Nevertheless, each approach has its own special advantages and disadvantages.

As mentioned, the synthetic approach is practically limited to proteins of about fifty amino acids or less, and in fact most work with synthetic peptide antigens has used pieces of eight to twenty amino acids. Intact protein antigens, on the other hand, are usually one hundred to two thousand amino acids long. In nature, these long protein chains assume a highly distinctive and predictable but complex and contorted folding. In the resultant three-dimensional shape, amino acids far separated in the sequential array may in fact lie quite close to each other in space. It was thus predicted that most antigenic sites would be conformational, i.e. requiring the whole protein to display its full shape. Nevertheless, short peptides, particularly those corresponding to the surface of the protein, can be injected and sometimes cause the production of antibody reactive with the whole intact molecule. For example, a chemically synthesized peptide twenty amino acids in length, coming from the immunologically most important section of a particular antigen of the foot-and-mouth disease virus, was capable of protecting guinea pigs against the whole virulent virus. In fact, on a weight-for-weight basis, this peptide worked much better than the whole protein of 213 amino acids length from which it came. Similarly, short peptides from influenza or hepatitis virus antigens have caused good antibody formation. While much work remains to be done to see how general this finding will be, recently-devised approaches to *predict* the immunologically most important portions of an antigen through a thorough study of its shape are looking interesting. On the whole, though, the peptide vaccine approach has not progressed as well as hoped. Part of the reason for this may be the fact that these short peptides are not as good as they should be in provoking the 'helper' T cells of the immune system. Furthermore, there is one biological constraint that has not received enough attention. Micro-organisms show a great ability to change and evolve. If a vaccine is directed against just one tiny portion of one antigenic molecule, there is a real risk that the microbe concerned will mutate in such a manner as to change that one component of its make-up, and thereby create a 'vaccine resistant' strain capable of eluding the host immune response, in much the same way that widespread use of antibiotics favours the emergence of antibiotic-resistant strains. To combat this possibility, synthetic vaccines should probably be cocktails of several different peptides, so that multiple separate mutations would be required to achieve immunoresistance.

A second disadvantage of synthetic vaccines, and indeed pure protein vaccines of any sort, relates to their strength as antigens. Living or killed micro-organisms frequently present antigens to the immune system as a bristling array of hundreds or thousands of molecules packed closely together on the surface of the microbe. This, for technical reasons which need not detain us, increases the intensity of the immune response. Furthermore, the micro-particulate nature of micro-organisms makes them palatable to the body's scavenger cells, and scavenger cell-associated antigen is a much more powerful trigger to the immune system than soluble antigen. In experimental situations, these disadvantages are over-come by the use of powerful stimulants of the immune system, called **adjuvants,** which are given with the synthetic vaccine. Most adjuvants are not suitable for human use because of toxicity and side effects. Several new adjuvants suitable for human use have been developed in both academia and industry. We will see which of these turns out to be most suitable for humans. Other approaches under continuing investigation include old-fashioned ones such as adsorbing the synthetic vaccine on to aluminium hydroxide particles ('alum precipitation') to achieve the needed molecular array and also a slow release effect; or newer methods of coupling of the synthetic peptide on to a 'carrier' molecule which is itself a strong antigen. Research aimed at strengthening immune responses deserves to be promoted, as it is common to *all* synthetic vaccines and indeed to the recombinant DNA approach as well. Unless the adjuvant problem is solved, there will be significant difficulties preventing synthetic vaccines being fully efficacious.

A third disadvantage of synthetic vaccines may be their cost, which presently is well above that of genetically engineered proteins. It is probable that costs will come down sharply as production technology improves. The major advantages of synthetic vaccines relate to their precision as chemical entities. There should be a minimum of batch variation and of unwanted side effects due to molecules not germane to the desired immune response.

There has only been one large trial of a peptide-based vaccine (otherwise called a Phase III trial—a trial in which people in endemic areas are immunised, allowed to be naturally infected and the outcome observed), namely the malaria vaccine of Dr Manuel Patarroyo, and unfortunately this proved unsuccessful. So currently this whole area is not very active within the vaccine industry.

The genetic engineering approach can make proteins of essen-tially any length, although most of the proteins that have been successfully made so far are less than a thousand amino acids long. Theoretical problems of finding the best part of an antigen molecule are thereby largely avoided, although it may still be wise to use a cocktail of different molecules to accommodate the immunoresistance problem. Genetically engineered vaccines need not be confined to one protein—it is possible to insert several genes and have them function in *E. coli*, thus making the bacteria into factories for ready-made cocktails of antigens. The Cetus Corporation has marketed a vaccine against scours, a toxic diarrhoea of swine, based on this principle. The Genentech group have produced a foot-and-mouth disease vaccine, which works in cattle, through genetically engineering the viral protein VP1. Even vaccines based on these full length proteins would benefit from improved adjuvants, so the search for effective adjuvants is doubly important.

A further, and somewhat avant-garde, advantage of genetic engineering is that potentially the DNA coding for the relevant antigens can be engineered into a living microbe which could actually grow inside the host being immunized, thus making a genetically engineered live, attenuated vaccine with all the attendant advantages of dosage and duration of antigenic stimulation. For example, Dr Bernard Moss at the United States National Institutes of Health has successfully engineered the cowpox virus, the very agent responsible for the global eradication of smallpox, to act as a carrier for several entirely different antigens. Dr Enzo Paoletti of the firm Virogenetics has modified this approach to use canary pox as a carrier, a very safe vaccine as this virus cannot multiply in the human host. Harmless gut micro-organisms can also be engineered to carry non-toxic antigens of intestinal pathogens such as cholera or typhoid, as we shall see. This is an active and exciting area of current research. Because the engineered organisms grow easily, genetically engineered vaccines will probably be inexpensive, except, of course, for the need to amortize research and development costs. Truth to say, however, the results in humans have so far been a little disappointing and progress much slower than anticipated.

The chief disadvantages of genetically engineered vaccines do not apply to vaccines dependent on living, engineered microbes but to those consisting of pure antigen molecules. First, the need to purify the antigen from all the other products made by the engineered organism, and, secondly, the question of antigenic strength, remain problems requiring attention.

While much of this discussion has focused on *E. coli* as a factory for pure protein antigens, and on living harmless microbes as gene recipients, there are many variations on these themes. For example, yeasts are frequently used as host cells in genetic engineering, not only because they can be grown so easily, but also because they are evolutionally closer to vertebrates than *E. coli*, and thus have the capacity to add sugars to some genetically engineered antigens which are mixtures of amino acids and sugars. Generally, yeasts synthesize and process proteins in a form that more nearly approximates what happens in human cells.

Animal cells are also being engineered successfully. While they are much more fastidious in their growth requirements, any description of the 'state of the art' technology would be remiss in not pointing them out as possible factories of the future. However, the greater cost of growing animal cells constitutes a problem for practical vaccine manufacture.

There are two further methods of immunization which are both still experimental and have not yet reached clinical use. They are worthy of mention because the first of these can do away with the requirement for better adjuvants and the second enables the production of cheap vaccines. The first new method is to use DNA coding for one or more protective antigens as the vaccine material. Such nucleic acid or 'naked DNA' vaccines have created great excitement. What happens is that the DNA of interest is embedded in a plasmid just behind a strong promoter and is injected into a muscle or into the superficial layers of the skin, and some of the DNA is taken up in the nucleus of the cells concerned, translated into messenger RNA and transcribed into protein. The antigen of interest leaks out of the cell, travels to the nearest lymph nodes and there induces antibodies and cellular immunity. Thus the body itself becomes a vaccine factory! Frequently the amount of antigen made is rather small, so tricks have to be used to ensure enough of it gets to where it has to go. Because of the stability of DNA, which does not require refrigeration, the relative ease of manufacture and the capacity to mix many plamids creating a cocktail capable of immunizing against several diseases simultaneously, DNA vaccines might be particularly useful in the developing countries.

The final and at first sight fanciful, still more experimental vaccine is an edible vaccine. This involves genetically engineering edible plants to produce important antigens together with an immune response-strengthening substance known as a mucosal adjuvant. The first small trial of this approach in human volunteers took place in 1997 and was surprisingly successful.

Vaccines in the pipeline: the challenges and the constraints

Given the above technological leaps, it is no wonder that academics all over the world are excited about all kinds of new vaccines or improvements in old ones. Dreams of great daring are being dreamt, extending the concept of vaccination from viruses and bacteria to single-celled or multicellular parasites and even to non-infectious diseases like cancer and multiple sclerosis. A birth control vaccine is the subject of active research. The sky seems to be the limit.

Yet, great though the need and the opportunity undoubtedly are, many academics underestimate the constraints which will ensure that new vaccines for human use will only materialize gradually. The first relates to funding. Vaccine research is expensive and risky, because research and development costs are high; but profits are likely to be low, because directly or indirectly governments are the major users of vaccines, and they are good at negotiating minimal prices. Moreover, drugs are used by patients daily for long periods, whereas once a person has been vaccinated, he or she only requires boosters at rare intervals, so the volume of sales is inherently lower than that of drugs. Human vaccines are less profitable investments for the pharmaceutical industry than drugs, and this is even more the case for those vaccines required chiefly in the developing countries.

The second constraint relates to the changing perceptions of regulatory agencies. Pasteur's rabies vaccine or even Jenner's smallpox vaccine would have great difficulties in today's regulatory climate, and indeed even the first tentative clinical trials would have trouble receiving approval by relevant ethics committees! It is, of course, essential to ensure safety, particularly as vaccines are administered to healthy individuals, but the risks of *not* deploying potentially effective agents rarely enter into the equation. Even if this issue is engaged for pure molecular vaccines, and is resolved, the difficulties with respect to suitable adjuvants and any living, genetically engineered organism as a carrier for antigens, will remain substantial.

The third constraint relates to expertise in the development component of research and development. Even though academics are buzzing with bright ideas about new vaccines, their capacity to translate a research breakthrough into a marketable product is notoriously limited, and partnerships with industry will be difficult to forge in this traditionally low-profit arena. Will academics have the patience to see a vaccine through to the development phase, and to conduct the extensive clinical trials that will be needed? This is much less heady work than the original genetic engineering, but just as essential.

It is appropriate now to consider some of the examples of vaccines that have appeared and those that appear to be within reach.

Hepatitis B vaccine: the first anti-cancer vaccine in history

Hepatitis means the liver is inflamed. There are many different causes but the commonest is infection with one of several viruses that attack liver cells. The hepatitis B virus is spread by exposure to contaminated blood, and infection with this virus is common in developing countries. A small proportion of people, for reasons that are far from clear, become chronic carriers of the hepatitis B virus and have large amounts of the antigen HBsAg in their blood. As many as 10^{13} particles (10 million million) can be present per millilitre of blood plasma. It is possible to bleed donors in such a manner as to remove large amounts of the fluid (plasma) component of the blood, but to return the white and red blood cells. Further, the HBsAg can then be purified from donated plasma and sterilized. In 1980 a clinical trial proved the capacity of this human-derived HBsAg to act as an effective vaccine, capable of preventing hepatitis B infection. In 1982 two firms, Merck and Co., USA, and the Institut Pasteur, Paris, independently marketed rather similar vaccines. This vaccine is 95% effective in its primary purpose, namely to prevent hepatitis B. Groups at special risk include people in frequent contact with blood products, such as physicians, nurses, workers in blood banks, laboratory personnel and dentists, and groups such as homosexuals and drug addicts. In many countries hepatitis B is now included for routine infant immunization.

Primary cancer of the liver is uncommon in Europe or America but is one of the commonest fatal cancers in Asia and Africa. Excellent evidence exists incriminating the hepatitis B virus as at least one of the causative agents of liver cancer, another major cause being hepatitis C. The *relative risk* of contracting liver cancer between chronic carriers and non-carriers is in fact higher than the relative risk of lung cancer in heavy cigarette smokers versus non-smokers, being 100:1 for example, amongst Chinese in Taiwan. A pathological sequence can readily be identified from viral destruction of liver tissue, attempts by the liver cell to divide rapidly to make up the damage, and finally frank liver cell cancer. It is evident from epidemiological studies that this progression takes several years. Though the details of how the virus causes cancer are not yet clear, one clue is that the genes of the hepatitis B virus integrate into the malignant liver cell. While the mechanism represents a great research challenge, the practical implications are evident right now. Logic suggests that the prevention of hepatitis B virus infection would prevent the eventual development of liver cancer, and this has turned out to be the case. One problem is that, in many cases, the hepatitis B carrier status develops in very early life, in the children of chronic carriers, through exposure to maternal blood during the birth process or via breast milk. Thus the vaccine must be given very early in life, or else babies will have to be protected by gamma globulin injected at birth and given vaccine some months later. Injecting very young children with the vaccine in Taiwan has already been shown to result in a dramatic reduction in liver cancer in the relevant cohorts of children.

Material from blood donors is not ideal as a source of antigen. Each batch of the vaccine must be carefully processed and extensively tested to ensure it carries no infectious agents. Several million doses of the human-derived vaccine have already been distributed, but the thought of vaccinating every child born into the world with human carrier-derived material strains credulity. Therefore, a genetically engineered vaccine is desirable both because of safety considerations and to enable large amounts of vaccine to be made. When yeast cells are engineered with the gene for HBsAg, they produce particles very similar to those found in the serum of human carriers, showing that the yeast cell can produce viral particles more or less in the way the human liver cell does. These yeast-derived particles have now replaced the human-derived vaccines in many countries and are proving to be every bit as effective. Costs were initially very high but have come rocketing down as mass production technology and competition enter the scene.

Vaccines against diarrhoeal diseases

Overall, the diarrhoeal diseases are as important to world health as the parasitic diseases. Perhaps most publicity has been given to cholera, because of its frequently dramatic manifestations and its capacity to cause brisk epidemics, but other causative agents are of even greater public health importance. These include the *Salmonella* infections typhoid and paratyphoid; *Shigella* infection (bacillary dysentery); infestation with amoebae (amoebic dysentery); and a wide variety of intestinal viruses. Diarrhoeal disease can interact with malnutrition and so an infection which might be readily controlled in industrialized countries may prove fatal in the sanitary and nutritional conditions of some developing countries. Oral rehydration has been an incredibly simple but extremely effective way of combating many diarrhoeal diseases. However, it will be many decades until environmental sanitation and personal hygiene practices in some tropical countries reach an adequate standard. Till then, other control measures will be required. The vaccine approach, with its capacity to prevent rather than cure, has enormous potential in this field.

The first generation vaccines against the major enteric diseases, for example those against cholera and typhoid, leave much to be desired. The injectable killed typhoid vaccines cause adverse side reactions and the protection conferred is only about 50 to 70%. The injectable cholera vaccine is also of low efficacy (50 to 70%) and its effects of short duration (six months or less). Fortunately, research was able to come up with some alternatives and these original vaccines are now obsolete.

On the typhoid front, an oral live attenuated vaccine developed in Switzerland is an improvement. This vaccine, termed Ty21a, makes use of a stable double mutant of the typhoid bacillus which has lost the capacity to make some of the enzymes required for virulence. The safety and efficacy of this vaccine was the subject of a three-year field trial in Egypt involving over 32000 children. No harmful side effects were noted, and even minor adverse reactions were scarcely above those of placebo controls. Over a three-year period, one case of typhoid fever occurred in 16 486 immunized children versus twenty-two in 25 628 unimmunized children. This 96% efficacy was most impressive. The vaccine is now in widespread use in many countries around the world and represents a significant advance in terms of safety, ease of administration and efficacy. It is also a good reminder that conventional genetics can sometimes be as effective as genetic engineering.

In cholera, great efforts to produce a better vaccine by bioengineering are under way. An orally administered vaccine based on killed cholera organisms has been tested and found to be effective. A variation on the theme is the oral coadministration of killed cholera organisms and a mucosal adjuvant known as CTB. One drawback is the requirement for at least two doses, a logistic consideration not always achievable in developing countries. Live vaccines have an advantage as they can often be given as a single dose. There have been several attempts to develop a live cholera vaccine. A live, attenuated cholera strain which goes by the picturesque name of 'Texas Star' was tried in normal human volunteers. It protected against subsequent challenge with live virulent cholera organisms. This strain lacked the gene for one part of the cholera toxin and thus did not cause disease. Its only drawback was that it caused mild to moderate transient diarrhoea in 24% of the volunteers, putting mass population administration in doubt. Subsequently other workers have developed strains of cholera that lack the toxin genes. These strains do appear to be free of side effects but unfortunately a large clinical trial in Indonesia of one of these failed to show vaccine efficacy. Another strand of research seeks to insert cholera genes into harmless gut bacteria by recombinant DNA technology. One variant of this approach towards an oral vaccine is based on introducing the genes for cholera antigens into the Ty21a typhoid strain, and, hey presto!—one has two for the price of one, a combined cholera–typhoid vaccine! This concept is progressing steadily through the clinical research pipeline.

Fabulous progress has been made in other areas as well. The meningitis vaccine known as Hib has proven effective far beyond early expectations, and the principles behind it will shortly result in similar vaccines against other forms of meningitis. The pneumonia vaccine against the so-called pneumococcus passed its clinical trails with flying colours and is proceeding toward widespread availability. An important cause of diarrhoea in infants is the rotavirus, where a new vaccine has recently been licensed, although there is currently some debate about possible side effects. An excellent injectable hepatitis A vaccine has been developed, although a live attenuated vaccine would also be desirable. Lists, if exhaustive, are also exhausting, and so we will refrain from summarizing exciting work in other areas. Suffice it to say that the above analysis is exemplary only. Vaccine research is alive and well for bacterial infections such as tuberculosis and leprosy, protozoal diseases such as malaria, sexually transmitted diseases such as syphilis, gonorrhoea and herpes, virus diseases like dengue fever, influenza, Japanese B encephalitis, as well as for a variety of special situations relevant more to developed countries, such as vaccines against gram-negative bacteria which cause infection in surgical wounds. We have not considered the many active projects investigating veterinary vaccines aimed at protecting economically important animals from disease. The more the power to manipulate microbes and antigens grows due to the continuing biotechnology revolution, the faster will these efforts come to fruition.

A blueprint for future action

'Health for all by the year 2000' was the stated goal of The World Health Organization. Now we are well past that date, and we can see that universal health will not be achieved by the six present childhood vaccines alone. They represent a solid beginning for a global immunization programme, but right from the start, the world should be thinking about the power of new vaccines. It is thought-provoking to witness at close quarters what two relatively modest (in financial terms) initiatives have done for research into parasite vaccines: the WHO/UNDP/World Bank Special Programme for Research and Training in Tropical Diseases, and the Rockefeller Foundation Great Neglected Diseases Program. Because of superb planning, selection of the most worthwhile lines of endeavour and the most able scientists, and a conscious effort to engage the minds and spirits of world leaders of research as active supporters of the initiatives, a catalytic avalanche has started, which is essentially unstoppable. The initial funding has been multiplied many times over as pressure on national and international funding agencies to join the fray has mounted. The conscience of the world is ready to be stirred by this cause. Many of the new vaccines have been waiting in the wings for too long, being largely the dreams of selected, small groups of scientists with limited financial and moral backing. The climate is changing; what was needed was a crystal in the super-saturated solution.

In 1998, the crystal appeared, from a surprising source. The world's richest man, Microsoft's chairman Bill Gates, together with his wife, entered the fray. At the time of writing, he has committed over US$1400 million to various aspects of global immunization. One of the authors has the awesome responsibility of being Chairman of the Strategic Advisory Council of the Bill and Melinda Gates Children's Vaccine Program. At the same time, a Global Alliance for Vaccines and Immunization (GAVI) has been formed, uniting the major United Nations agencies (especially WHO, UNICEF and the World Bank) with private foundations, non-government organizations, development aid agencies, academic institutions and vaccine manufacturers into a broad but focused coalition. GAVI has four linked aims. First, it seeks to renew and revitalize the Expanded Programme on Immunization, which seems to have plateaued at a coverage of 80% of the world's children. This figure should go up, particularly as the global statistic hides a great heterogeneity, with some African countries reporting less than 40% coverage. Secondly, the global eradication of poliomyelitis is envisaged early in the new millennium. Thirdly, the newer vaccines must be added promptly and in sensible, prioritized order. Fourthly, research into new and improved vaccines must be further intensified. For example, a vaccine against HIV/AIDS is the only realistic hope for controlling that disaster in the developing countries. In sub-Saharan Africa, an additional four million cases are being infected each year and in many countries 20% or even more of the adult population are HIV-positive. Malaria and tuberculosis are other major killers requiring new vaccines.

The new international dynamic has instilled a surge of hope. Realizing that time will be needed, the authors believe that as GAVI gets into its stride, it has the potential to at least halve the nine million deaths per year occurring in children from infectious diseases. GAVI could represent the greatest public health triumph in history.

9

Genetically manipulated organisms— new plants, animals and bacteria

The implications of genetic engineering for medical research and practice have received wide attention, not only in the media but also within the scientific community. Great though the medical challenges are, many believe they will eventually be overshadowed by a wide diversity of uses of genetically engineered species in primary, secondary and tertiary industry; genes will become technological slaves. Indeed, it is possible to dream large dreams; anything from vastly improved crop yields to replicating biochips as a revolutionary approach to manufacture of integrated circuits has been the subject of speculation. We have discussed the methodology that allows genes to be identified and transferred into new host organisms. This is being routinely done for micro-organisms to turn them into producers of new proteins (Chapter 5). Nothing in the technology prevents the gene recipient from being a higher organism such as a plant or an animal. Bacteria, plants or animals bearing new genes are often referred to as genetically manipulated organisms (GMOs). In this chapter we shall look at some of the issues surrounding GMOs. Bear in mind that, even with a global enthusiasm for this stunning technology, the pathway from research bench to mass markets is thorny and tortuous, particularly where legitimate environmental and regulatory concerns enter the picture. Previous experience with revolutionary technology has led the general public to be somewhat wary of the mixture of benefits and problems that seems to invariably ensue.

Genetic engineering in agriculture

Genetic manipulation of crops is as old as civilization, forming the basis of agriculture ever since the practice spread out from the

Fertile Crescent. Any selection of seed varieties from amongst a range of available options is, in fact, a genetic experiment. Until this century astute empiricism guided the choices; but since World War II plant genetics has become a major profession with a regular parade of triumphs. The 'green revolution' which has allowed global food production to keep up with an expanding population rests on a disciplined and institutionalized, ever-changing research base. For example, the Consultative Group for International Agricultural Research, a consortium led by international aid agencies, United Nations agencies and private foundations, runs a group of research institutes devoted to improving food production, and its annual budget is over US$330 million. The development of new plants was based on conventional genetics, namely experimental mating and repeated selection for desired characteristics. This was in turn supported by sophisticated plant physiology, knowledge of the structure and function of plants which guided the principles on which selection was based. The scientists selectively mated plants with desirable characteristics. They also made astute use of mutations in which errors in DNA copying occasionally, by chance, threw up a plant better suited to a particular environment. One criticism that has been levelled at conventional agricultural genetics is that many new crop varieties have demanded more intensive use of fertilizers, which have risen drastically in price over the last decades and in some cases cause environmental pollution. The question is whether genetic engineering can do this job faster and better, allowing previously unimagined hybrid or mutated species to come forward with extraordinarily desirable characteristics.

Moving genes into plants

Genetic engineers have spent the last twenty-five years amassing an arsenal of techniques that allow them to move genes around at will. Some plants have an extraordinary advantage over animals for the would-be genetic engineer. The so-called dicotyledonous plants can grow from a single cell, so that, in theory, all the scientist has to do is to introduce the desired gene into that one cell, and a genetically altered whole individual could be grown. Things are not so simple for most of the economically important plants. They belong to the so-called monocotyledonous variety, and these

is still an enormous amount of work to be done to understand what these genes do and how they interact in the living organism. Genome projects have commenced for wheat, rice, maize, sorghum, alfalfa, kidney bean and rape oilseed. Moreover, if we remember that DNA is a universal code, then if we find a gene in bacteria or in other species that does something useful, there is no reason why we cannot introduce this into a plant, so that the plant gains the same useful property.

We can illustrate this concept with two particular examples. Let us begin with a modified form of corn that is currently being developed in research laboratories. All proteins are made of amino acids, but animals including man are unable to make all the amino acids necessary to make proteins. They must obtain these by digestion of proteins in food. Lysine in particular is an essential amino acid, but it is often in low quantities in vegetable sources such as corn and soy bean. So, the production of corn rich in lysine would provide not only a useful human food, but also a highly improved animal feed. Now the enzymes for making lysine in corn are inefficient, in the sense that as soon as any lysine is produced, two of the enzymes stop working. This feedback inhibition stops the accumulation of lysine and keeps levels low in corn. We could attack this problem in two ways. We could redesign these enzymes in corn so that this inhibition no longer occurs (though that sort of fine tinkering with protein function is still beyond our capabilities). Or we could use bacteria in which these same enzymes are not inhibited by the build-up of lysine and go on working to make more lysine. Taking the bacterial genes and placing them into corn under the control of appropriate promoters, will if all goes well and the genes are controlled properly, result in corn kernels with a higher level of lysine. It will also mean that now we have a corn plant with two novel genes, originating in bacteria.

The second example concerns a new form of rice that has already been constructed and is available for use. Vitamin A is an important nutrient for humans, being involved not only in the efficient working of the immune system, but also being essential for sight. Now rice is an important subsistence crop in many parts of the world, but the rice kernel that people typically eat is very low in Vitamin A. The result of this widespread vitamin deficiency is hundreds of thousands of cases of blindness as well as deaths in children unable to resist particular infectious diseases. One way to solve this problem is to supply these people with Vitamin A supplements, but no organization has stepped forward to do this logistically difficult distribution. An alternative is to engineer rice so that the kernels become rich in Vitamin A. This would require introducing multiple genes that encode a number of enzymes within the Vitamin A biosynthetic pathway. Scientists working in Switzerland and Germany, Drs Ingo Potykus and Peter Beyer, have done just this, taking genes from a bacterium and the daffodil and introducing them into rice. The genes in fact produce a precursor of Vitamin A called beta-carotene which is turned into Vitamin A in the human body. Beta-carotene is a yellowy-red colour and the resulting rice has been nicknamed 'golden rice'.

So now we have methods of introducing new genes into plant cells and a means of producing new plants. We have techniques for manipulating many of the major food crops and in the next few sections we will ask some questions about what we might wish to achieve, review what new varieties are available and consider some distant prospects.

Insecticide and herbicide resistant plants to feed a hungry world

Probably the greatest problem facing humanity is the increasing numbers of people on this planet. So many other problems, including environmental degradation and widespread pollution, loss of other species and habitats, war, poverty and malnutrition result from this situation. How can we manage to do even such a simple thing as provide food for these people? One approach would be to increase global production of food to meet this increased demand. However, this will have to be done more efficiently and that requires improvements to existing techniques. At our current levels of production, significantly more of the earth's area would have to be converted to croplands to feed the planet's growing population. This would have further serious effects on animal and plant populations in the wild and would decrease the amount of global species diversity.

Most experts believe that the conventional methods of improvement in yield of food crops have gone about as far as they can. There have been impressive achievements, such as the doubling of yield in the USA over the last 40 years. Although crop yields are still increasing by approximately 1 per cent per year, the rate of increase has slowed markedly. Further improvements in yield, crop quality and crop protection will need new technology to keep pace with population increases. This is an argument regularly advanced by proponents of GMO development and it is quite compelling. If we have exhausted conventional technology, what is there with the potential to keep pace with the increasing demand for agricultural products and also protect the environment and ensure the safety, security and diversity of the food supply? Nothing other than GMOs appears able to do this. Opponents to GMOs argue that such reasoning ignores other approaches to the problem such as the regulation of the world population, something we will have to do sooner or later. The rate of population increase is levelling off and the problem does not appear as acute as it did even a few short years ago. The reasons for this are complex and include increasing prosperity in Asia with consequent decreases in family size. There is also the demographic impact of the tragic AIDS epidemic in Africa with its negative effect on population increases. So the magnitude of the need may not be as great as suggested. They would also argue that much could be gained by improving current methods of food distribution and storage, two processes in which massive amounts of food are currently lost or spoiled. If these problems could be addressed, then the food needs of the world are not likely to increase so quickly. These are good points and should not be ignored within the debate on GMOs. Nevertheless, the possibility that food could be produced more cheaply, at a higher quality and with less impact on the environment is highly attractive.

What are companies currently doing with GMOs? The leader in this area is undoubtedly the Monsanto Corporation. It has a number of products currently available for planting in which introduced genes have conferred new properties. Their products at present fall into two general groups: a group of plants that are resistant to the herbicide Roundup and a group of plants that make their own insecticide. Roundup is a non-selective herbicide that essentially kills all plants. However GMOs have been modified to produce an enzyme that protects plants containing it from Roundup. Thus farmers can spray their crops during the growing period. The weeds are killed and the GMOs survive. According to the company, this will reduce overall herbicide use and remove the need to use long-acting herbicides that leave a residue. Crops already modified in this way include soy beans, corn, canola and cotton.

Insect damage is a major cause of loss of production. There are a number of insecticides available with differing properties. One of the commercially used insecticides is a purified protein from the bacterium *Bacillus thuringiensis*, or Bt. Bt creates a toxin that is indigestible to target insect pests, and kills the insect by accumulating in the digestive system and damaging it. Studies indicate that Bt toxin is harmless to people and other living things, including many beneficial insects that can help control other pests. Bt toxin has been available for at least thirty years and has been fairly extensively used. Plants are modified so that they make the Bt toxin by incorporation of the bacterial genes into the plant DNA. Thus the leaves contain the toxin and it has the same effect as if it had been applied to the leaves by external spraying. The advantages are that it only affects insects feeding on the leaves, and insecticide no longer needs to be sprayed around. These seem useful outcomes and a number of plants including cotton, corn and potatoes have been modified with Bt toxin or other proteins that specifically damage certain insect pests. It is of course possible to combine these introduced properties and produce, for example, corn that is both herbicide-resistant and produces its own insecticide. As we shall see, there has been considerable controversy about the conduct of Monsanto in producing plants whose benefits are linked to the use of Monsanto products.

Genetic engineering and nitrogen fixation

Other forms of modified plants are more distant prospects. Proteins, the nutritionally most important component of food, consist chiefly of carbon, hydrogen, oxygen and nitrogen. The availability of nitrogen in soils is one major factor limiting the productivity of agricultural land. Extensive use is now made of chemical fertilizers containing nitrogen, for example in the form of nitrate or ammonia. As nitrogen is present in such large amounts in the air that we breathe, why is its availability a problem? To be useful to plants or animals, the nitrogen has to be in a form where cells can readily use it and place it appropriately in proteins. The vital conversion of atmospheric nitrogen gas into potential organic building blocks is called biological nitrogen fixation. Only very lowly organisms, bacteria and blue-green algae, have evolved the extraordinary trick of converting nitrogen gas into the ammonium ion. However, certain green plants have done something equally clever. They have entered into a symbiotic contract with nitrogen-fixing bacteria, in which the bacteria actually supply little factories of ammoniacal fertilizer for the plant. Legumes such as soy beans, clover and alfalfa use this method. Next time you have the chance, examine the roots of one of these plants. You will see small rounded nodules which are, in fact, the homes of the nitrogen-fixing bacteria and the producers of the plant's own supply of fertilizer. In contrast, important cereal grains such as wheat and corn lack these nodules.

The biological production of nitrogenous fertilizers is under the control of a set of genes called Nif genes. Major research thrusts are aimed at the better understanding and control of these Nif genes. The twin purposes are to improve the efficiency of already existing nitrogen fixation, and to devise ways of allowing plants like wheat to fix their own nitrogen. It has already been possible to create strains of *Klebsiella* bacteria that have a mutation which causes them to keep on producing ammonium ions even while abnormally large amounts are accumulating in the environment. The mutant strain has shut off the normal feedback mechanism. Experiments with these organisms have shown that biological nitrogen fixation is extremely energy intensive, and a search is under way for the most efficient species that could harness solar energy for manufacture of fertilizer. One novel research project is seeking to improve an old, traditional three-way symbiosis. Blue-green algae help a tiny water fern, *Azolla*, to grow in rice paddies, and as the *Azolla* decays, the fixed nitrogen becomes available to the rice plant. Genetic engineering is seeking to induce the *Azolla*–algae system to export increased levels of fixed nitrogen, thus finally increasing rice yield.

Super plants to thrive in all conditions

There are other properties that might be desirable in the next generation of plants. A significant problem in many countries is soil degradation and loss of arable land. For example, important agricultural regions in the USA and Australia are being turned into salt deserts as salt levels in the soil rise. Plants vary greatly in their capacity to tolerate increased salinity. Obviously seaweed and plants that grow by the sea are highly tolerant, whereas wheat and citrus fruits are very sensitive. Salt tolerance is under genetic control and work is going on to identify the genes that control salt-resistance as a prelude to moving them into important crops.

Another major agricultural problem is the destruction of plants by insect pests. The widespread use of insecticides is a much criticized but often necessary means to prevent the depredations of insects. But there are other ways. One approach we have already considered is the introduction of the Bt gene, and this is effective against some insect pests. How might we extend the range of such natural insecticides, and what could we do if pests developed resistance to Bt? Plants evolving in a world where insects have always been present have developed their own means of combating insects. Amazing as it may seem there are substances made by plants that kill insects yet are harmless to humans. One protein with anti-pest properties is trypsin inhibitor. This protein, as its name suggests, inhibits the action of trypsin, a digestive enzyme produced by many insects that feed on plants. The insect trypsin digests the plant cells into a form that the insect eats, but by feeding the insect produces holes in the leaves or stems of the plant. The trypsin inhibitor stops all this and the insect thus deprived of its food eventually dies of starvation. Scientists have so far successfully introduced the cowpea gene for trypsin inhibitor into tobacco plants. When budworms and army worms were placed on the transgenic plants, both types of insects failed to grow and eventually died. Trypsin inhibitor is broken down in the human digestive tract and is harmless to humans. It is known that a number of major plant pests, the rice ant worm, the corn ear worm and the boll weevil, are sensitive to trypsin inhibitor and introduction of this gene into plants such as rice, corn and cotton could have a major impact on improving yields of these economically important crops.

Clever though such ideas are, it is not yet possible to know which of these manipulations will eventually lead to improved crops. An intelligent policy for agricultural research for the next few years should allow plenty of scope for the fundamental researcher to investigate the structure and organization of plant genes

at a basic level. As in all aspects of research and develop
detailed fundamental knowledge will itself suggest new p
approaches. Would that our policy-makers could learn and
to terms with this deep truth!

Genetically manipulated animals

Much of the development of GMOs has occurred in plants and
the controversies about its use mainly focus on crop plants. This is
because the methods available to scale up plants to the amounts
required for food production are very efficient. Thus once a
transgenic plant has been constructed, industry can relatively
quickly produce large amounts of seeds. Monsanto has bought a
number of seed companies to ensure that it has the capacity to
produce seeds relatively quickly. In contrast, the time taken to
breed a single transgenic sheep to the hundreds of thousands or
millions required for food and wool production is very much
longer. A single sheep might give rise to only one or two lambs per
year, which in turn must reach sexual maturity. Indeed, it is likely
we will see the widescale adoption of in vitro fertilization and
cloning techniques to accelerate this process. Thus commercial
exploitation of GMOs will initially focus on plants, though there
are a number of interesting developments to do with animals.

As with plant breeding, our manipulation of animal breeding
reaches back into prehistory. Over the last few centuries this
manipulation became more systematic as prize-winning animals
were mated to improve the bloodline. The limitation of such an
approach has always been, as for plant breeding, that it is somewhat
haphazard: you may want one trait such as rapid weight gain but
also get another such as too much fat on the animal. Additionally
there are inherent limits within a breed beyond which it is not
possible to go. However, just as with plants, it is possible to cut and
paste genes and introduce them into pigs, sheep, goats and cattle.
These GMOs are also often called transgenic animals and the new
genes they contain are called transgenes. There are several ways to
produce transgenic animals and we will describe one technique. A
fertilized egg is immobilized on a microscope stage (this egg has
usually been fertilized in the test tube by the combination of eggs
and sperm obtained from animals). Using the microscope and a

paratus to hold it, a very fine injection pipette
into it are injected hundreds of copies of the
[15]only about 1% of eggs survive this treatment and
divide the transgene is passed on to every cell in the
includes the animal's reproductive cells so that the
can be passed on to successive generations. A major
—exactly as for gene therapy (discussed in Chapter 7)—is
gene may be successfully put into the cell but it might not
ion. As we mentioned in Chapter 2 it is regulatory sequences
rounding the gene that govern its expression, and these must be
chosen carefully so that the gene can be coaxed to function but
only at the right time and in the right tissues.

The world's first transgenic animals were made by Ralph Brinster
and Richard Palmiter in the United States. They took the gene for
growth hormone, a hormone normally made in the brain, and
hooked it to a regulatory sequence from the liver that controlled a
gene for binding metal ions. By feeding the animals on zinc they
switched on the transgenic growth hormone (in the liver) and
produced mice that were twice as large as normal. The implications
for increased meat yield in farm animals were obvious and transgenic pigs carrying human growth hormone were constructed.
Unfortunately none of the pigs grew faster and they had several
abnormalities such as arthritis and decreased fertility. This sort of
result demonstrates that proteins are part of complex regulatory
networks and the uncontrolled expression of one sort of protein
may cause unexpected consequences. Further work is often needed
to redesign the genes to avoid these problems.

An alternative means of achieving the same ends of increased
production with growth hormone is not to engineer the animal but
to treat it with its own growth hormone, which has been produced
in bacteria by standard gene shuffling techniques described in
Chapters 3 and 5. This approach is commonly applied in the
United States in cattle using bovine growth hormone (also known
as bovine somatotropin). This product, produced by Monsanto
and called Posilac, has been available since about 1994. It is
currently used in about one-third of all dairy cows in the USA,
and is the largest selling dairy animal health product in the United
States. Its use results in increased milk production, in the order of
two to seven kilograms per day per cow, leading to increased

profitability for the producer. A number of organizations including the United States Food and Drug Administration, the World Health Organization, the Food and Agriculture Organization and regulatory agencies in over thirty countries have examined the milk produced from cows treated with Posilac. They have concluded that there are no particular concerns and that milk produced in such cows is indistinguishable from milk in untreated cows. In particular there are no detectable residues of the hormone in the milk. Yet despite this, there remains a great deal of unease about the use of such products, particularly in the case of beef cattle treated with the same hormone.

Because of the difficulties in scale-up of production of transgenic animals, in the short term it is likely such animals will exist only in small numbers. However, they might still have a major economic value if they are used to produce products of high worth such as human pharmaceuticals. The feasibility of this has been shown many times in a laboratory setting. A simple example might be the production of therapeutic substances such as Factor VIII (used to treat haemophiliacs) in the milk of transgenic animals. As we mentioned in Chapter 5, it is often better to make such substances in mammalian cells to ensure that sugar molecules are attached correctly. The gene for the clotting protein is attached to a regulatory sequence that normally controls milk production. Such a construct only functions during lactation and the protein that is made is secreted into the milk, where it can be readily collected by normal milking procedures. It can then be purified from other constituents of the milk. The company that has taken this approach furthest is Genzyme Transgenics, based in the United States, which focuses on the production of human therapeutic proteins in the milk of transgenic animals, predominantly goats. This is because of the combination of large milk volumes, relatively short time to breeding in each generation, and ease of handling and milking. It would not require very many animals to produce amounts of protein sufficient to treat tens of thousands of patients. Their general approach is to first demonstrate the protein can be expressed in mice and then to shift it into goats. They have expressed more than 50 different proteins in this way including antibodies, serum components and proteins for use in vaccines, and commercial products are on the way.

What might some other uses of transgenic animals be? One area of active research is the possibility of developing animals that could serve as organ donors for people. When organ transplants using kidneys or hearts are performed, there is a scrupulous need to match the tissue types of the donor and recipient, otherwise rejection of the transplanted organ occurs. We have mentioned tissue types in Chapter 6. There is a critical shortage of donor organs in most countries and many patients die of their diseases before a suitable donor can be found. Animals have similar organs to humans but currently cannot be transplanted because the rejection process is even more severe against animal organs. The most acute provokers of rejection are certain sugar molecules found on the animal cells. Research is focusing on the genes coding for the enzymes that make these distinctive sugars. If these genes were removed then one barrier to successful cross-species transplants, called xenotransplantation, would be gone. Additionally the animals could be engineered so that their cells expressed the same tissue type antigens as humans. There could be a line of animals for each of the major tissue types. At present the favoured animal for these experiments is the pig as the physiology of this animal is very similar to that of humans. There are several ethical and technical concerns that must be addressed including the possibility that transplantation may introduce some as yet unrecognized animal virus into the recipient, but this research continues apace.

Other prospects are more distant. Scientists at Australia's CSIRO are looking at ways to increase wool production. Often sheep are limited in the amount of wool they make by lack of an amino acid called cysteine. Sheep cannot synthesize this amino acid but must get it from the food they eat. Other organisms such as bacteria can make cysteine from another amino acid already present in the body of the sheep. The gene coding for these cysteine-making proteins has been cloned and is being transferred into sheep.

As if farm animals were not enough, molecular biologists have turned their attention to fish. Fish have several features that make them very easy to work with. Fertilization of eggs occurs outside the body and large amounts of eggs and sperm can be readily obtained. The eggs are large and can be readily microinjected, and are easy to maintain after fertilization, developing in water, outside the body of the fish. Of course all the usual problems of introducing a gene successfully into the nucleus and having it expressed at appropriate times in useful amounts, are present. New genes have been successfully introduced into rainbow trout, goldfish and several different types of carp. Some obvious applications include the introduction of genes that make fish grow faster or bigger, make them disease resistant or make them able to digest new types of food such as byproducts of agricultural processes. Such transgenic fish may be of immense use in helping to feed mankind. It seems likely that transgenic fish would be maintained in special fish farms and would not be released into natural ecosystems, though the possibility of accidental escape would have to be kept in mind in every instance.

Genetic engineering and the production of naturally-occurring microbial products

It is easy to forget that micro-organisms already produce a large variety of products that are of industrial value. As we have been discussing agriculture, products of use to the food industry come to mind. Vitamins such as Vitamin C, Vitamin D, Vitamin E and nicotinic acid can all be made by biotechnology, and while their consumption by human populations is not likely to rise greatly, they could play a much greater role in stock feeding if they became sufficiently cheap. Amino acids constitute a business approaching US$2000 million per year, being widely used as nutritional supplements and flavouring agents. Japan has a high profile in this industry and an advanced biotechnological research capability in the field. Nevertheless, processes for the large-scale manufacture of nutritionally essential amino acids remain too expensive for routine addition to stock feed. At present, annual world sales of two of these, lysine and methionine, total US$500 million. An examination of trends over the last decade clearly shows that sales would rise considerably if unit costs came down. We have discussed one way to approach this already by redesigning plants, but an alternative would be production in bacteria and addition to the feedstocks. This may have attractions if the opposition to the use of genetically modified plants becomes widespread. Enzymes are also vital to the food industry, being used in literally dozens of processes from cheese production to meat tenderization. These natural products of bacteria—vitamins, amino acids and enzymes—represent just three of many examples where the rate, efficiency and finally cost of production could be progressively and greatly improved by the thoughtful manipulation of the genes of the organisms used. Again, industry experts are predicting that the scene will have changed totally within five to ten years. It could be argued that conventional genetics, too, could have been used for such purposes, but somehow the power and appeal of the new technology has brought a whole series of new forces in its wake. There is a new enthusiasm about industrial microbiology which is worldwide, not confined to the academic community, and bound to impinge on production technology, though to what degree it is hard to foretell.

Genetic engineering in production technology aimed at high-volume products

The pharmaceuticals and biologicals which are useful in medicine represent specialty products of high value manufactured in relatively small amounts. Amino acids and enzymes are medium volume/medium value products. However, there are also some products required in very large volume, but of low value per unit volume, that have been widely discussed as subjects for advanced gene technology. Perhaps the most interesting of these is alcohol.

Ordinary alcohol, or ethanol, has great potential as a transport fuel. It can be mixed with petrol one part to ten to form 'gasohol', which can be used in unmodified automobile engines. Relatively small changes to engines would allow much larger proportions of ethanol. Apart from that, the world already uses several billions of litres of industrial ethanol each year, as an industrial solvent and in many other uses. This alcohol is made by fermentation followed by refining. Essentially, three approaches to cost reduction are under intensive research. First, more efficient organisms are required for the fermentation. Imagine the potential of a yeast strain that drove the alcohol content of a fermented fluid from the present maximum of 13 to 14 per cent to, say, 20 per cent or even higher. This would be exciting enough; but what if the yeast could drive the various chemical reactions not at room temperature, or even at blood heat, but say at 60°C? These are the sorts of goals ahead of genetic engineering technology. The second factor in the equation is to find cheaper raw materials to begin the fermentation, including wood wastes, crop residues such as wheat straw, or wastes from processing of sugar cane. Here, also, novel organisms capable of tough and unusual bioconversions would be required. Thirdly, imagination would have to be used to devise cheaper refining methods.

Obviously, there are too many unknowns in this equation for anyone to be confident in making relevant economic predictions. A proposition that is frequently bandied about is that, if oil were to go to US$50 per barrel, ethanol (or methanol) could become economically feasible as an alternative source of transport fuel. Of course, at this oil price, oil shale or coal liquefaction also look attractive, and it seems unlikely that oil prices would go to these levels in the world as we know it at present. But cataclysms in the Middle East could influence the situation, as could major research discoveries in biotechnology.

Many other products in this general category could be listed. Ethylene glycol and ethylene oxide between them constitute a market of some US$10 billion per year. Currently these are made chemically, but they could be made by fermentation, and the same applies to propylene. Production of methane gas from wastes could serve as an energy source, and of course in its turn fermentation of wastes could prove to be a major boon for waste disposal. Quite a few biotechnology firms are already active in the waste disposal field. Hydrogen has been spoken of as another energy source which could be derived from industry waste materials through appropriate microbial action.

Attractive though the above examples may be from a research viewpoint, in practical terms most of them seem a long way from the marketplace. Almost certainly, what we shall see is not a blinding rush of new process technology but a gradual seepage of the new thinking into the industrial scene. Some of the glittering prizes eventually to be won certainly justify continued research.

Genetic engineering and the mining industry

It is amazing how perceptions of the mining industry have changed over the last few decades. Thirty years ago mining was seen as the coming boom industry in many exporting countries, a welcome passport to increased affluence. Then came a new emphasis on environmental concerns and a realization that the earth's resources were finite, finding expression in publications such as the Club of Rome's *Limits to Growth*. Further developments of exploration technology and advances in geophysics and geochemistry soon led to the next phase, namely a recognition of vast new mineral deposits, frequently of low grade but of immense size. This, in turn, evoked an interest in cost-effective recovery methods including obvious things like economies of scale but also in greatly improved technology. For most of the earth's crust's resources, it now appears that the supply will last a very long time, particularly if extraction continues to become more efficient.

New recovery techniques are relevant not just to recently discovered ore deposits, but also to existing and even abandoned mines. With conventional mining technology, samples of rock released by explosion are analysed to determine metal content. Technological and economic considerations determine what percentage of metal is required to warrant further processing. If a given area of a mine yields a result below the calculated cut-off grade, that particular lot of rock is carted away as waste. The next stage of mining might involve a concentration step, where finely crushed rock is further sorted into the bits with much metal and those with less. Both waste and tailings still contain far more metal than a random sample of rock.

With both low-grade deposits and wastes, the question comes up as to whether the metal content can be won in some way that makes economic sense. Here is where some see a bright future for genetically engineered micro-organisms. Microbes can sometimes grow and thrive in the most surprising conditions. The aphorism 'nature abhors a vacuum' is true for biology as well as physics. In biological environments that seem hostile to most forms of life, organisms with strange characteristics can often be found, products of the long Darwinian struggle. So there are micro-organisms that love metal-containing ores, and moreover in their normal metabolism, they often leach the metal from the rock within which it is buried. Biotechnological recovery of uranium or copper through bacterial action is already a reality, and often the metal is actually converted to a soluble form in the process. For example, insoluble metal sulphides are converted into metal sulphates. In principle, these procedures could be applied either to low-grade deposits or to mining wastes.

Some of the bacterial reactions capable of extracting metal require an external energy source. Here, a speculative possibility is that two micro-organisms working together might be the techno-logical slaves, one capable of harnessing solar energy through the process of photosynthesis, the other with the enzymic machinery for bioleaching. It seems probable that, provided sufficient time and effort were expended on study of the fundamental chemistry and biology involved in these processes, genetic engineers could tailor-make organisms that were better than naturally occurring ones in these bizarre activities. Indeed, recent research on bacteria capable of leaching nickel from lateritic ores looks distinctly encouraging. However, a number of considerations suggest that practical applications are still in the more distant future. Most metals are in oversupply at present, and available at low prices through current technology. Legitimate environmental concerns about release of the micro-organisms on a large scale would need to be addressed and deployment would probably have to be in a phased manner. The mining industry as a whole is much more attuned to physical and chemical technology than to biology. None of these reasons, however, should act as a barrier to continued long-range research.

Slaves to fit any need?

Some areas of current research have received scant attention in this chapter. A great deal has been done in the area of food processing, for example cheesemaking, brewing, baking and the wine industry. These and the related field of fragrances and flavours are already dependent on biotechnology and clearly capable of further improvement. The chemical industry has grave problems as the major polluter of large rivers. Biodegradation processes could remove toxic wastes from water before its discharge, and could help in other areas of waste disposal. The chemical industry is also constantly on the lookout for more efficient methods to achieve its chemical conversions, where special enzymes could come to be regarded as efficient catalysts. The oil industry is excited about organisms capable of dealing with oil spills and others aiding in oil recovery. Biological control of agricultural pests is an enormous area where true progress has been made. The shopping list is increasing monthly, not through trivial or fanciful additions, but through a permeation of ideas and excitement into the most unlikely areas.

A frequent criticism of genetic engineering is that the release of transgenic plants and animals will cause environmental catastrophe. Perhaps we should conclude this section with a look at how genetic engineering can help repair ravaged areas of the environment. One of the most disturbing aspects of large-scale industry has been generation of substantial amounts of toxic waste substances. The plastics industry, for example, uses special chemicals that enable basic ingredients to combine and form diverse types of plastics. These chemicals are synthetic, i.e. they do not occur in the natural environment and are made up of unusual combinations of atoms. Polychlorinated biphenyls (PCBs) are made up of rings of carbon atoms to which are attached chlorine atoms. This attachment is very unusual in living organisms and thus micro-organisms have not evolved ways to digest these compounds. Thus PCBs and related substances such as dioxin (Agent Orange), chlorinated phenols and benzenes are not biodegradable (a biodegradable substance is one that is digested or broken down by organisms in the environment such as bacteria, moulds or algae) and persist for very long periods of time. In addition to this long life these substances are poisonous as they interfere with normal living processes. Thus, dioxin was used as a herbicide because of these poisonous properties.

Large amounts of these substances are stored in metal drums in toxic waste dumps throughout all industrialized countries. However they have also escaped into the environment through leakage, deliberate dumping or incorrect handling practices. Occasionally news stories on places such as Love Canal in the northeast USA alert the public's attention to this continuing problem. Intensive studies on naturally-occurring bacteria have identified several bacteria that are capable of digesting these environmental toxins. For example there is one bacterium that can remove the chlorine atom from PCBs, a key step in degrading these compounds. Other organisms can degrade TCE and TCA, major contaminants of ground water in areas of the USA. Other researchers are designing bacteria in the laboratory, selecting mutant genes that degrade toxins or engineering new metabolic pathways. Most of these organisms do their job slowly and do not completely remove the toxins. An experiment performed by Ananda Chakrabarty in the USA is particularly interesting. Dr Chakrabarty had identified a bacterium with the capacity to break down a chlorinated hydro-carbon. He grew the bacteria under conditions that increased this ability. Eventually he was able to develop a strain of bacteria that used these compounds for food. He added these bacteria to soil samples that were contaminated with dioxin (Agent Orange). The level of dioxin decreased from 1000 to 7 parts per million within one week. He showed that seeds planted in the soil could now grow. As with many examples in biotechnology, once the initial principle has been demonstrated, the fine tuning can proceed rapidly. One major problem that will have to be addressed is whether such organisms (either engineered or naturally selected mutants) will be allowed to be released into the environment in large amounts.

Research potential lies at one pole, commercial needs and realities at the other, and in between sits a puzzled and frightened society. We will now consider the issues engendered by this un-comfortable equation.

Introducing new species: corn and potatoes or rabbits and prickly pears

To conclude this chapter let us discuss one of the most difficult issues in the application of genetic engineering. The power and elegance of genetic techniques are unquestioned, but should we apply them outside the laboratory or the health clinic? There are risks and benefits associated with the introduction of new tech-nologies and genetic engineering is no different in this respect. Each new use will have expected benefits and risks, both those that are foreseeable and those that are unpredictable. The desirability of proceeding then needs to be assessed based on an estimation of the likelihood of benefits and risks and their probability of occur-ring. For some uses, such as the production of new pharma-ceuticals, the answer is a resounding yes. We can see great benefits to the production of TPA for example, with very little risk to the factory worker, society or the environment. But there are uses where the benefits may not be so great or the risks greater or more difficult to predict. Is this the case with transgenic organisms? In a surprisingly few years, we have passed from a theoretical discussion of the release of transgenic organisms to the situation where vast acreages of land are currently being used for cultivation of transgenic plants. How did this rapid transformation come about and how did we know it was safe to proceed?

To illustrate the factors involved in making such decisions, let us consider how the release of transgenic plants into the environment came about. This is undoubtedly one of the most controversial applications of DNA technology, and it is extraordinary how far it has progressed in the USA. How could we tell what would happen if we took a modified organism created in the laboratory and then released it into the environment? Was this an irresponsible gamble by greedy companies or was there some rationale to it? Are we likely to wreak ecological havoc as the engineered organism spreads widely, destroying naturally occurring species? Could the transgene make its way into natural populations of plants and animals with unforseeable results? Have we made a catastrophic mistake in allowing the external release of altered organisms?

From one point of view we are repeating history in that the early history of recombinant DNA technology was marked by similar controversy. Conceivably, genetically altered organisms, created in the laboratory, could infect the research worker or escape into the environment. In Chapter 11 we will review the history of recombinant DNA and see that a solution to this danger was the physical containment procedures and the creation of enfeebled bacterial strains. Subsequent experience has led us to believe that these precautions have rendered these experiments safe, and certainly there is no documented evidence of any hazards after over two and a half decades of such experiments.

But the release of engineered organisms is different and we cannot rely on physical or biological containment. Indeed, these organisms are being released because we wish them to multiply and thrive in the environment. How could we have predicted what would happen when the different corn and soy bean varieties were planted? In fact there was a considerable amount of data on which to draw. This data had been accumulated from the study of the introduction of natural species into new geographical areas. Examples included the importation of animal and plant species by settlers or by returning explorers. Some of these importations were highly beneficial. One need only list the importation of food plants to Europe from the Americas in the fifteenth and sixteenth centuries—corn, potatoes and tomatoes for example. Other importations had bad consequences with the imported species becoming a pest, for example the introduction of the rabbit into Australia. By studying a large number of examples of such importations, scientist were able to arrive at a number of criteria that should be met in order to minimize the risks of an engineered organism spreading too widely.

The experience of studying these 'experiments of nature' is that in general the probability of the introduced organism spreading out of control is small, but if it does the damage could be quite extensive. For example, of foreign plants introduced into Australia less than 10% became established and 2% spread widely enough to be thought of as weeds. Similarly in Britain about 1% of introduced species have become pests.

The criteria that warn us that a new species may spread out of control include such properties as rapidity of reproduction, and how rapidly it moves around the local environment, ease of killing the organism, on how wide a scale it is released and perhaps most importantly, how adaptable it is. If an organism can survive and thrive in a very large range of conditions then it is likely to find large areas where it can grow. Thus an organism that divided every twenty minutes and could grow happily at many different temperatures, utilizing many different food sources and was resistant to poisons would be very likely to spread widely. In contrast, an organism that multiplied once every twenty years and was extremely fastidious in its growth requirements might survive in a limited area or die out completely. Therefore the genetic engineer needs, as far as possible, to tailor the new organism to fit a very specific niche in the ecology. Additionally it would be prudent to contain the transgenic organisms if possible, for example transgenic fish could be grown in fish farms rather than released into the ocean.

Such considerations can only provide guidelines because we do not yet know enough about the ecology to predict with total accuracy the outcome of a release. Experience had to be gained by experiments in which slightly modified organisms were released, tagged in such a way that their spread could be monitored. In this way a body of knowledge was amassed that allowed better prediction of outcome. With such knowledge we were able to consider the release of more commercially important species. Each release could then be considered on a case-by-case basis with clearly defined criteria of danger and constant monitoring so that release could be terminated if required. An essential prerequisite is the close study of test plantings where the ability of the plant to spread or the transgene to move into new populations is carefully followed. If these experiments are successful, then the way is open for progressive release of the organism, but the process may take years. In many cases it is possible to gain further comfort from the behaviour of the original plant itself. We understand the properties of corn very well, and it highly unlikely that adding a single gene is likely to make the corn spread widely.

So the release of transgenics has proceeded apace. In 1988 over a dozen novel organisms were released in small-scale field tests. Most were plants but new strains of bacteria were also released. One was a bacterium that was sprayed on to potato plants as its presence inhibited the formation of ice crystals and subsequent damage to the potato crops. A great deal of public controversy attended those tests. It was clear that the companies had not spent sufficient time consulting with and educating the public. Matters were further exacerbated when it was discovered that a researcher had disobeyed US federal rules when releasing bacteria designed to protect trees from Dutch elm disease. Such maverick behaviour must be roundly condemned. The public has the right to expect responsible behaviour from scientists and companies.

However, notwithstanding these problems, these early releases paved the way for the introduction of a number of commercial transgenic plants. Monsanto released its first transgenic plant for commercial planting in 1996. Growers in the USA enthusiastically accepted these new varieties and 3 million acres of transgenic plants were planted in that first year. In 1997, additional varieties were made available and these resulted in plantings of 19 million acres. This was a phenomenal rate of growth, and it was estimated by 1999 that 33% of the corn crop and 50% of the soy bean crop were now provided by transgenic plants. We can at least be sure that such crops are not becoming plant pests, and are remaining confined to the area of planting. However, are they safe for humans and the environment?

GMOs and the environmental movement

On this issue a major battle is being fought right around the world. The response to the use of GMOs differs markedly between the USA and Europe in particular. For whatever reasons, American consumers initially accepted the widespread use of GMOs in the food supply without serious qualms. The very rapidity with which this change took place may have surprised many. In recent times, there are more concerns being expressed within the USA, and this appears to be in response to the widespread opposition in Europe. The opposition within Europe is passionate, ranging from the destruction of test plantings of GM plants, to the refusal of major supermarket chains to carry products from GMOs, to a host of reports and recommendations by governments and august bodies.

Although GMOs have the potential to benefit consumers in many ways, in general they have been received extremely negatively by the environmental movement. Greenpeace has led a consistent campaign against genetic engineering in its various forms and against GMOs in particular. However opposition is not restricted to this single group and there are organizations, both public and governmental, that have been extremely skeptical of such developments. In some countries such as Germany where the Green Movement is represented in the legislature, a number of regulations have been enacted to restrict genetic engineering in various ways. In Switzerland some cantons voted on a proposal to ban various types of genetic engineering experiments. The proposals were ultimately defeated, partly because the large Swiss pharmaceutical companies warned that they would be forced to relocate to other countries if their research and development efforts were to be hampered in such a way.

There are a number of reasons why environmental organizations, particularly in Europe, have taken this stand. One relates to the recent tragic history of bovine spongiform encephalopathy or mad cow disease, where unsafe agricultural practices have apparently resulted in the spread to a few humans of a fatal neurological disease. Although this had nothing to do with genetic engineering and transgenic animals, it sensitized the public to issues of food safety and how decisions made on economic grounds in the production of food could adversely affect human health. Activists became very concerned that the official agencies were not examining this new technology closely enough but were being manipulated by commercial interests. Secondly, there is a spectrum of moral and animal rights concerns, arguing that the introduction of foreign genes is perverting the essence of the animal and constitutes a sort of eugenics. In the view of some, humanity simply does not have the right to meddle with other species in this unnatural way. This has particular resonance to many Germans, because of past policies. Thirdly, there are the potentially unpredictable effects of movement of the transgene into natural populations and its possible effect on other plant species, or insect populations. A recent study suggesting that pollen from Bt-toxin producing plants might kill larvae of the monarch butterfly fuelled such concerns. Further, there have been worries that some introduced proteins may act as allergens in a small percentage of the population, causing sensitivity reactions. Genetically manipulated plants with the capacity to grow in previously inhospitable conditions might increase the acreage of land under cultivation and put pressure on wild species of plants and animals that previously lived there. Biodiversity would in this scenario be further compromised. Finally, there are issues to do with globalization and harmonizing world trade practices. Some of the GMOs such as the laurel rapeseed will produce oils that render unnecessary the purchase of palm and coconut oils currently produced in Third World countries. Such policies may lead to further impoverishment of developing countries. In addition, trade agreements might prevent countries banning some products they feel are unsafe, if they have been accepted in countries such as the USA.

Mad cow disease was an unpredictable result of a practice adopted to save money in the provision of feed to cattle. Could food from GMOs be another example? How were such organisms being assessed?

Monitoring the development of GMOs

A question of concern to many is 'How stringently is this new technology being examined for possible problems?' Once introduced, products from transgenic animals are likely to be consumed by tens of millions if not billions of humans. How extensively does one need to test to assess side effects? As scientists practising in health care related fields, we are aware of several medicines that passed initial safety and toxicity testing in clinical trials. However once released onto the market, serious side effects were encountered in the larger populations of patients who then took the drug. Now we do not know if there will be serious side effects from GMOs, and it could be argued that if these occur in a very small number of people, it should not prevent the deployment of important new crops. As with most aspects of life, there is a trade-off between benefits and hazards, and a decision is made to continue where the benefit is overwhelming.

However there is one major difference between GMO-derived food and medicines being prescribed for an illness. Food is being given to those who are healthy, and for whom many other alternatives to these foods are available. In the case of patients, they are experiencing an illness that will have a deleterious effect if left untreated. It could be argued therefore that the safety standards need to be correspondingly higher for food.

Is the new technology being stringently assessed? Europe and the USA have taken different approaches to the regulation of GMOs. Within the USA the approach to GMO products is determined within the Food and Drug Administration and the Environmental Protection Agency and to some extent the Department of Agriculture. Their view has been that foodstuffs have constantly been altered over time, whether by conventional breeding techniques, recombinant DNA or by food processing methods. Foods are an incredibly complicated mix of tens of thousands of substances, the net effect of which is beneficial, although individual components such as fats or toxicants may have some damaging properties to humans. Foods such as cassava can be quite poisonous if not prepared properly. The addition of a single gene making a new protein that is not known to be harmful, will in the view of the FDA do little to alter the foodstuff. Thus if the food is 'substantially similar' to a pre-existing food then there should be few regulatory problems. It might be that if changes are made to the nutritional content, such as varying the composition of fats, specific labelling would be needed. For example, the herbicide-resistant soy bean contains a gene encoding a protein similar to one already present in soy bean. The difference is that this gene comes from bacteria and encodes an enzyme that is resistant to herbicide, whereas the enzyme normally present in soy bean is sensitive. Thus the difference is a few amino acids in a single protein out of tens of thousands. The FDA views this as substantially similar and the transgenic plant is approved for use. In contrast, the European view has been that products resulting from biotechnology are novel and they require novel regulatory oversight. They would argue that here is a plant with a new piece of DNA never likely to have been in plants and this makes it substantially different. How these plants should be regulated in Europe is still being widely debated, and Europeans have adopted the view that the press of events should not be allowed to overtake this process. Thus until the method of regulation of these products is determined, the products should not be allowed on the market. This is an ongoing process, and the outcome is not clear. Clearly, each example needs to be examined on a case-by-case basis, and we need to amass a body of data which will inform us as to which is the more reasonable point of view.

Where is such data coming from? In the mid-1990s a series of public conferences called the Risk Assessment Research Symposia were held. The proceedings of those conferences are still accessible on the Internet and make for interesting if highly technical reading. Reports on a number of transgenic plants and microbes are presented. They show that in certain circumstances such crops have a selective advantage over non-transgenics and would be likely to spread. It is also clearly shown that transgenics will mate with normal plants and that the transgenes will be passed onto that generation of hybrid plants. We are sure that even proponents of the use of GMOs would accept such findings, as they are what one would expect. The question is what significance do such findings have for plotting the spread of such plants and genes from the original planting sites. Strangely, these symposia appear no longer to be held and the results of such experiments are now presented in other forums.

Consumers can have a major say in how this new technology will be deployed. By demanding that labelling be more informative and that it accurately list the origin of the ingredients, they can be in a position to make an informed choice. Pressure can be applied to supermarket chains to ensure that they have a clear policy on the sale of products from GMOs. Already in Japan and Europe some supermarkets refuse to stock foods derived from GMOs. This could result in increased prices because of the extra work required to achieve this end, but many might view the action as worthwhile. This has major implications for food-exporting nations. It could therefore be that in the short to medium term a country could gain a marketing advantage by studiously avoiding any kind of GMO, thus projecting a 'clean, green' image. In the longer term, it is likely that some GMOs will have such clear advantages as to be irresistible, particularly in a world that will have to feed 9 to 11 billion people by 2050. As with the original controversy about genetic engineering, the conjectural hazards will have to be examined on a case-by-case basis. Two levels of examination are necessary, namely a rational and informed debate, and a staged experimental deployment looking for unexpected effects in suitable contained environments. Whether this is occurring in a context where transgenic plants go from 0% to 50% of the US soy bean crop in a little over four years is debatable. Having said that, it is difficult to identify any ill effects of this major change in food producing habit. We must maintain a vigilant attitude and appropriate continuing monitoring practices must be put in place.

As for our personal view, we would argue that this technology has enormous benefits but it can be developed at a studied and leisurely pace. The extraordinary expansion in acreage planted with these varieties has probably occurred too rapidly. Is there an imperative to proceed at this pace? There is no GM food on the market that currently provides a major benefit to the consumer, that is unavailable from alternatively produced foods. Perhaps the 'golden rice' variety that is rich in Vitamin A is the first of such improved foods. Thus for now, most benefits are to the grower. However for companies to pursue further improvements, there must be a realistic expectation that they can recover their research and development costs from these products and that they should be profitable. Companies with an eye on their balance sheet and stock prices will want to hasten the deployment of these products, but is there a need? After all, we live in a world where many countries destroy stockpiles of food to maintain produce prices. We probably have a reasonable amount of time in which we can start developing the new plants required to survive in regions of excess salinity and degraded soil. The current species of transgenic plants offer some benefits, but in our view none of these are a compelling reason for such widespread adoption so quickly. We would also argue that more stringent controls should be applied for transgenic organisms with the capacity to spread widely, including bacteria and fish. At the same time we need to look carefully at the practices that have led to environmental damage. Maintaining the status quo by an increasingly complex series of technological fixes is a practice that will ultimately fail. The application of these techniques will buy us some time but their use must be part of a comprehensive re-evaluation of many of our environmental policies and uses.

10

The DNA industry

An obligatory partnership links science and technology with industry and commerce. The partnership began with the industrial phase after World War II. It has changed the face of society with breathtaking speed, and in many ways has determined the agenda for humanity for this century and beyond. For example, perhaps the most thought-provoking issue in Australian domestic politics, at present, is the correct way to adapt to technological change, and to face a future in which microchips rather than people will bear incremental work loads. So Silicon Valley's brainwaves have taken only a decade to become dominant global concerns.

Within the free enterprise system, new technology is brought to the marketplace through the activities of corporations, frequently helped by government initiatives of various sorts. The pathway from innovative idea to saleable product can take a variety of forms. In the case of the largest companies in innovation-intensive industries, their own research and development divisions may be so large and competent that the whole process goes on within the one corporation. More often, the academic sector in universities and research institutes plays the dominant role in fashioning new concepts, and passes the baton for practical development. On other occasions, individual inventors or small companies drive a discovery a certain distance and then seek the help of large companies for further development. But in the case of the DNA industry, it appears that a new genre of activity has been born, its patterns a sharp departure from past practice and its *modus operandi* an exciting, if unproved, harbinger of future direction.

The hallmarks of the DNA industry are threefold. Brains are unashamedly the chief asset; research results, be it a genetically engineered organism, a new gene, or a novel biological, are the chief products; and a speculative form of capital raising, termed venture capital, the chief method for financing operations. In this chapter we shall look at the history of this young industry and examine both its virtues and its problems.

The birth of the DNA industry

The 1940s saw a great upsurge in research on the structure, function and genetic nature of micro-organisms. Ever since Pasteur realized that fermentation of beer or wine was due to microbial action, it has been clear that microbes are useful chemical factories, but until the 1940s art and craft had been more prominent that science in guiding the relevant process technology. Ernst Chain and Howard Florey's work on Alexander Fleming's discovery, penicillin, changed all that. They showed that penicillin was a wonder drug in the treatment of bacterial infections, but production was hampered by the low yield obtained from the *Penicillium* mould. The first broths in which they grew their *Penicillium* yielded barely one part per million of the precious therapeutic substance. The reasons are not hard to determine. Micro-organisms normally make just as much of a particular substance as is useful to them. They are equipped with sophisticated metabolic control systems which prevent wasteful or excessive secretion of a particular metabolite. These control loops depend on genetic circuitry. In the early 1940s, after Florey took his moulds to the United States, seeking the help of that country's pharmaceutical industry to optimize production, scientists treated *Penicillium* with X-rays. This form of irradiation causes damage to the DNA double helix, and, in the aftermath, faulty repair occasionally supervenes. The slightly altered DNA may manifest itself as a mutant organism. If you X-irradiate a sufficient number of organisms, and have sufficiently powerful selection techniques, sooner or later you will find a mutant that has lost the control loop limiting penicillin production. That higher-yielding organism can then serve as the starting point for further genetic manipulation. This essentially simple strategy improved penicillin yields more than a thousandfold, with corresponding reductions in cost, rendering mass production feasible.

This success ensured that industrial microbiologists watched with great care the ever more intricate techniques being developed by academics for the elucidation of metabolic pathways and the detection of mutant organisms. Japanese industry was outstanding here, making major leaps in the technology of producing essential amino acids. Indeed, industrial scientists were so successful in harnessing mutation-based genetic manipulation that they soon reached the limits of what conventional genetics could do for them. Improved strains could do so much but no more; what was needed for the next quantum leap were entirely new strains with tailor-made characteristics.

The first company that recognized and articulated this need clearly was a small Californian group, the Cetus Corporation, which could be termed the founder of the new, elite club that constitutes the DNA industry. Cetus was founded in 1971 in Berkeley, California. Its early aim was to combine biological and engineering capability in a new way. Dr Donald Glaser, a Nobel laureate at the University of California, Berkeley, had invented a machine, irreverently termed 'the Dumb-waiter', which used modern electronic control systems to process up to 100 million microbial cultures simultaneously, permitting automatic surveillance and screenings, thereby greatly speeding up the search for mutant micro-organisms. Dr Ronald Cape (a PhD in biochemistry who had gone on to do a Master of Business Administration course at Harvard University) and Glaser founded Cetus, with Cape as President. The initial plan was to obtain contracts from the pharmaceutical industry to produce micro-organisms capable of increasing the yields of antibiotics. But then, in 1973, came the major revolution. Stanley Cohen's group at Stanford University and Herbert Boyer's group at the University of California, San Francisco, succeeded in constructing biologically functional DNA molecules that combined genetic information from two different sources. Genetic engineering was born. Cetus was ideally placed to enter the fray, as Stanley Cohen and Stanford Nobel laureate Joshua Lederberg, the pioneer of bacterial genetics, were amongst its early consultants. The 1975 report of Cetus Corporation makes interesting reading. It states :

We are proposing to create an entire new industry, with the ambitious aim of manufacturing a vast and important spectrum of wholly new nonmicrobial products using industrial micro-organisms . . . We propose, therefore, to transfer the genes for human interferon, in one program, and for human antibodies, in another program, into industrial micro-organisms, and to produce large quantities of these compounds in industrial fermentation. To our knowledge, no such programs are under way anywhere in the world . . . The opportunity defies adequate description.

Prophetic works indeed in 1975!

The DNA industry spreads

Cape and his colleagues were not alone in their perception that a new industry was being born. In 1976 Genentech was founded in San Francisco to exploit the hormone research started by Herbert Boyer, and, from a tiny base, this group has achieved many triumphs. In 1977 it announced the production of the first genetically engineered hormone, somatostatin. Boyer and many of his colleagues were then still university-based, which caused some problems. He later joined Genentech full time, since when genetically engineered human growth hormone, insulin, interferon and many other proteins have been made by this able group.

Another company early into the field was Biogen, associated with Charles Weissman of the University of Zurich and Walter Gilbert of Harvard University. They have done very good work in the interferon field, and have a ready-made marketing outlet for their products. In 1979 the founders sold 16 per cent of the stock to the large pharmaceutical firm, Schering-Plough, for US$8 million, creating the first such close association between a DNA company and a multinational drug house.

By 1980 the founding of companies based on recombinant DNA had come to resemble a gold rush in the United States. Though four groups, Cetus, Genentech, Biogen and Genex, occupied most of the headlines in the financial pages, many others had equally impeccable academic connections. Leading Harvard scientists Mark Ptashne and Tom Maniatis founded the Genetics Institute. Collaborative Research of Waltham, Massachusetts, was advised by Nobel laureate David Baltimore and Stanford geneticist Ronald Davis, among others. Bethesda Research Laboratories is strategically located close to the huge United States National Institutes of Health and is highly regarded by many scientists there. New England Biolabs was founded by Donald Comb of Harvard University. By 1983 there were 150 small companies in the United States based on genetic engineering and other advanced biotechnology. Since then, most of the leading molecular biologists in the United States now have some link with a company, ranging from consultancies to major shareholdings and directorships. Indeed, a major sport at US scientific meetings is to guess which scientist has formed what new company, and who will be inveigled to join it!

Though activity in other countries has been less frenetic, nearly all the major OECD nations are making investments in the field. In many countries the involvement of governments has been direct and supportive. Upgrading of the biotechnology industry is major, explicit government policy in the Federal Republic of Germany and in France. Canada, the United Kingdom, Switzerland, the Netherlands and Belgium are all prominent in the field. In the meantime, the interests of the developing countries have not been forgotten. The United Nations Industrial Development Organization, UNIDO, has worked hard to establish an International Centre for Genetic Engineering and Biotechnology which aims to speed technology transfer in the field, promote global cooperation and address problems of special relevance to developing countries. The UNIDO centre is at two sites, one in India, where there is a significant concentration of first-rate scientists, and the other in Italy. While there is still doubt in some quarters as to whether developing countries are yet in a position to contribute or to benefit from the genetic engineering revolution, the concept is worthwhile, because such a centre could help not only developing countries but also small developed countries, which currently find it difficult to enter this competitive field. As we saw in Chapter 8, recombinant DNA technology will be critical for some Third World health problems. Also, it is unlikely that Western companies will address these health problems. They have failed to do so in the past and the increasing cost of new drug development makes it unlikely that they will do so in the future.

Australia has made a significant effort to establish a biotechnology industry. Some twenty or thirty companies are active in the field. One outstandingly successful firm is Biota Holdings Ltd, which together with Glaxo-Wellcome has developed the drug Relenza to treat—and prevent—influenza. Other active firms include AMRAD, Biotechnology Australia, Circadian, Novogen, Progen and Florigene, and doubtless by the time this volume appears others will have come to prominence. Australia has a surprisingly distinguished record in basic biological science, and it would be a pity if all the bright ideas could only be developed overseas. One attractive feature of the DNA industry is that the capital required for entry is not enormously high. Brains and originality are what counts. However, large costs are incurred in the developmental stages of research and especially in the safety and efficacy testing of products designed for human use. For this reason linkages with an established multinational pharmaceutical company are frequently established.

Too much enthusiasm too soon?

There is no doubt that the prospects for the DNA industry have often been the subject of considerable hyperbole. Let us return to the United States pioneers and see how they initially financed the developmental research for which they became famous. One example is Genentech, which marketed the first major seller of the DNA industry, TPA, an anti-clotting agent. How did Genentech get the money to start? Back in 1977 Robert Swanson, the company's president, had raised nearly US$1 million for Genentech from sources such as International Nickel Company and the venture capital firm, Kleiner and Perkins. It seemed an enormous amount at the time! By 1980 the company had grown to the degree that a public offering of stock seemed appropriate. A million shares were offered at US$35 each, but the providers of this US$35 million ended up with only 13 per cent of the company! Yet, on the first day's trading, the shares shot up (briefly) to US$89! At that stage, Wall Street was theoretically valuing the company at over half a billion dollars and Herbert Boyer's original investment of a few hundred dollars had a paper value of US$80 million. For comparison, the Upjohn Company, one of the leading, established pharmaceutical firms, at that time had a market capitalization of US$1 billion. This excessive enthusiasm did not last long, and the Genentech stock settled back to its issue price, but huge theoretical values for start-up companies still occur. When Schering Plough bought 16 per cent of Biogen in 1979 this set a paper value for the group of US$50 million and by 1980 Biogen was looking for more capital on the basis of a self-assessed value of US$100 million. At the time, the firm was employing a total of sixteen scientists in its laboratory in Geneva, and had not generated any sales revenue! Yet twenty years later, the early confidence was richly vindicated, and the market capitalization is many billions of dollars.

Perhaps the scale of access to venture capital just discussed, and the degree of equity dilution which the later waves of investors appear prepared to accept, are peculiarly American characteristics. Nevertheless, a substantial amount of private capital has also been expended in other countries. DNA companies have gone bankrupt and it seems likely that many more will do so, a sequence of events that occurs in all commercial 'gold rushes'. The real question is whether those that do succeed will reap rewards for their share-holders commensurate with the risks that have been taken. Certainly the total world market for the products coming forward is potentially enormous, but these new specialty companies face several real problems. First, gene cloning technology is constantly improving and much of the new knowledge will remain in the public domain, making it possible for competitors to catch up quickly. Secondly, patents in this area will be difficult to defend. One example has been the successful challenge by Wellcome Biotechnology of the Genentech European patent for TPA. In other cases, companies have successfully protected their 'breadwinners'. Amgen, a biotechnology company founded in 1980 and now valued at over US$50 billion, has been able to maintain its patents on erythropoietin, despite extensive challenges. Thirdly, and perhaps most importantly, many new companies do not have either the experience in production technology and scale-up of larger, older companies or established skills in marketing. It seems inevitable that, in most cases, links of one sort of another will have to be developed with established larger corporations. With considerably less fuss, many of the leading drug firms, such as Eli Lilly or Hoffmann-La Roche, have added skills in genetic engineering to their own in-house research expertise, and their greater experience with the regulatory procedures imposed by governments before products can be sold gives them a major advantage. However, the best biotechnology firms will succeed and represent an important and beneficial adventure.

Biotechnology at the millennium

Having reviewed the history, let us examine the status of biotechnology at the dawn of the new millennium. A conservative estimate sees some 3000 companies of which perhaps 300 are major players. The most successful, Amgen, has two products each of which enjoys sales of over US$1 billion per annum. Amgen's success, a decade after Genetech went public, stimulated a speculative frenzy for a couple of years, but biotechnology firms have also had their lean periods, e.g. 1992–1994 and 1996–1998. In 1999, the market capitalization of the biotechnology industry in USA was US$150 billion. The top twenty companies accounted for two-thirds of this amount. Of the 365 publicly listed companies, two-thirds had a market capitalization of less than US$100 million.

The invention of the **polymerase chain reaction** enormously speeded up the search for new genes and the arrival of a 'first draft' of the human genome in 2001 has identified many more. New genes create new targets for drug discovery and new ways of looking at the biochemical abnormalities underlying disease processes. Functional genomics (Chapter 4)—the study of genes and how they act in health and disease—is combining with robotic technology and information science in a massive acceleration in the search for prevention, early diagnosis and more effective treatment. In the United States, for example, pharmaceuticals now account for sales of around US$100 billion, or 10% of total health costs. As biotechnology totally revolutionizes health care, this proportion could treble in the next fifteen to twenty years. Over the last twenty years, the industry has produced about twenty-five medical products with annual sales of around US$11 billion. About forty more products could reach the market within the next two years, a marked acceleration of success. As a matter of fact, for cancer treatment alone, 200 clinical trials are ongoing. Overall, the product range is very diverse: monoclonal antibodies; drugs which target new cellular receptors; drugs designed through detailed knowledge of protein structure; lead compounds discovered through an amalgam of genomics, combinatorial chemistry, high-throughput screening and superior data management. There seems little doubt that biotechnology is about to realize its early promise.

Of special interest is the nature of the relationship between the biotechnology companies and the major established players in pharmaceutical industry. A good start-up company will usually embark on joint ventures with several 'big pharma' partners, as the deep pockets help to pay for early stage research, and the development experience comes in useful when a lead needs to be tested in the clinic. Conversely, an astute research director of a big firm will comb the research endeavours of the start-ups, recognizing the energy and fleet-footedness of the small firms, thus effectively outsourcing some of the research. It can be a mutually beneficial relationship.

The biotechnology industry and the university sector

The emergence of the biotechnology industry has radically and irreversibly altered the relationships between the commercial world and the university world. There have, of course, been reasonably close relationships between these worlds in other branches of medical science. Departments of pharmacology in medical schools interact widely with manufacturers of ethical pharmaceuticals. On the whole, however, a rather subtle and not altogether worthy tension has surrounded this relationship. Frequently, the best medical scientists have ended up in universities and research institutes and have looked down somewhat on those of their colleagues who have gone to industry. Some academics have gone even further and regarded drug companies as little more than milking cows, good for providing money in their research, but otherwise of little worth. In departments of biochemistry, genetics, microbiology or immunology, knowledge of industry has been minimal and contacts confined to a few particular instances. What the new DNA firms did was to bring a new and much more free-wheeling atmosphere to the industrial research laboratory. Scientists working for these companies say that they do not feel confined; the ebb and flow of discussion and the looseness of administrative controls resemble the atmosphere of a university. This has proved attractive to many good young minds.

The stellar character of some of the entrepreneurs has also helped. If a Donald Glaser, a Herbert Boyer or a Walter Gilbert backed a venture, could it be anything other than exciting and worthwhile? The scientific advisory boards played their part—in fact, in some cases, they were even more luminary than the scientific advisory boards of the purest of the elite research institutes! Associates of the better firms started sending some of their best postdoctoral fellows into the DNA industry, not—as had been the case in the past—the weaker ones who would not quite 'make it' on the academic scene. So excellence became the norm and, as in the universities themselves, a self-perpetuating element. Scientists began to realize they could be rich *and* famous—this was something new and strange. It is worth adding that over the last decade as university finances have tightened, the established pharmaceutical industry itself is also attracting some excellent research stars.

Shifts in perception as basic as this do not occur without much trauma. Turmoil and unease has swept through the universities as a result of the birth of the DNA industry, and some of the worries are real. The chief ones relate to conflict of interest and secrecy. All over the world, academics draw their main support from government grants: in other words, from taxation dollars. In a real sense, therefore, the creativity of academics, who, though modestly paid are, in many instances, securely employed, should benefit the whole community. Yet when a breakthrough of potential importance to industry occurs, there is the temptation to derive some personal gain or benefit from it. Different institutions approach this problem in different ways. In Australia, for example, it was the norm to regard all patentable discoveries as belonging to the employing institution, which in fact took out the patent and derived the ultimate financial advantage. This was regarded as fair, because the institution had borne the costs and assumed the risks of the research. The institution itself frequently had to share its rights with a governmental granting agency. In the past, the inventor would gain indirectly through prestige, promotion and perhaps greater security of tenure. Now most institutions are adopting a policy of allowing some portion of the royalties to go to the researcher, typically perhaps 30% of the total. With a successful product, this can amount to a very large sum. In addition, some scientists augment their income through consulting. Even so, the level of remuneration to research scientists is very low compared to gifted individuals in other fields, and the pressures for a more commercial outlook are sure to increase.

In the United States a more entrepreneurial spirit exists, and many universities allow academics not only to retain a proportion of the patent rights of their discovery for themselves, but also to spend significant time on commercial work. This is seen as giving individual scientists more incentive to think about the eventual practical implications of their work. The universities take a very stern view of scientists who achieve a development while working as academics under a grant, and then promptly take their discovery, for example a genetically engineered clone, and join a venture capital firm. This has already been the subject of lawsuits in the United States, but abuse of the practice is very hard to police and doubtless there are many marginal cases where part of a discovery was made in a university laboratory and part after the move to industry.

There is great dissent over the question of whether a professor can work simultaneously for a university and directly within a venture capital firm (as distinct from consulting with it). Some universities regard this as perfectly acceptable, provided the university knows of, and has approved, the arrangement. Others forbid the practice and insist that the conflict of interest would then be too great—the scientist is forced to choose between his old loyalties and peer group and the new potential prizes. Again there are no clear-cut rights and wrongs in the situation; so much will depend on the cultural norms of the society concerned, and the personal integrity of the individuals.

There are other areas of conflict of interest that encompass somewhat subtle but nonetheless important concerns. Throughout this century researchers at universities have been free to study virtually any aspect within their discipline. The choice has been made on criteria of fundamental importance and interest. This concentration on pure research has laid the foundations on which the current revolution in molecular biology is built. Discoveries such as the isolation of restriction enzymes or of plasmids would never have come from an industrial R & D unit because practical uses for them were not immediately obvious. When industrial money is given to universities to fund research, it is often for medium to short-term projects which by their nature are goal-oriented, and aimed at production of products. Thus, there is a tendency to alter the direction of research efforts. Why struggle to attract money to study slime moulds when there is money available to work on cloning a blood clotting factor? Where a unit is well funded it becomes an attractive place to work in, because of its modern equipment and improved working conditions, and this will attract better staff and students from non-commercial, but perhaps fundamentally more important subjects of study. Thus, an impoverished underclass develops of scientists who are working on areas not at present believed to be commercially interesting, and science itself is redirected. Fortunately, it appears that many companies are wary of killing the goose that lays the golden egg, and offer their money to underwrite the activities of the department without dictating direction, but reserving first rights to any commercial spin-offs. One wonders, however, how willing they will be to renew that commitment if no products emerge after five years of funding?

Another area at conflict of interest concerns the role of the scientist as the impartial advisor to governmental bodies and regulatory agencies. We as a society will need to grapple with some monumental questions in the near future. The release into the environment of transgenic plants, animals and microbes, gene therapy experimentation on humans to reverse defects and the widespread use of genetic screening for disease prediction are just some of the weighty issues that we as a society must face. To make informed decisions on issues with profound impact on our way of life, our environment, and our view of ourselves, we must have the best available advice. With most of the world's top molecular biologists on the boards of biotechnology companies or consulting for them, will we receive the impartial advice we need? The industry as a whole is affected by regulatory decisions. When the US Supreme Court allowed the patenting of life forms, it meant a boost in stock prices for many companies. Likewise a moratorium on the use of transgenic animals and plants in agriculture would have a serious dampening effect on the market and indirectly the stock holdings or fees of most scientists. Of course, such conflicts of interest occur and are resolved ethically in many other areas of business and government and we have no reason to believe that it will happen otherwise when scientists confront the real world. However it underlines the realization that we as scientists are being forced to leave the ivory tower to take our place in society.

The second set of problems relates to commercial secrecy. A creeping secrecy within molecular biology represents a real threat. This discipline has a glorious tradition of openness, ideas flying across the oceans at great speed, and competition relating more to brilliance of insight and elegance of experimentation than to thoughts of financial reward. This tradition is beautifully described in Horace Judson's *The Eighth Day of Creation*. Yet it is being eroded, through the boom in genetic engineering. Scientists who were once so eager for their few minutes in the limelight within the peer group are turning coy at the large international conferences, skirting over details and even refusing point blank to reveal technical or structural data. This is destructive of progress in molecular biology which has depended not just on the sharing of concepts but also on trenchantly strict peer group criticism of techniques and results. A new sequence seen with increasing frequency is: discover, patent, call a press conference to boost your stock, then publish! Irksome though it may seem, this is still much better than the norm in some industries, where publication is delayed for months or even years for commercial reasons. Despite present trends, the death of the normal tradition of free exchange of information seems unlikely. The motivation to shine in the eyes of one's most distinguished colleagues is still very strong amongst scientists, and good lectures at conferences plus first-class written papers are still the best ways of achieving that peer group esteem which is so treasured.

An orderly future for the DNA industry

It seems reasonable to predict a more orderly, if pluralistic, development of the DNA industry in the third millennium. There is no doubt that the fundamental technology of gene cloning, expression, manipulation and analysis will continue to develop with amazing speed and to dominate biomedical science over this period. However, workers in the DNA industry will not just be recipients of this technology; they will increasingly contribute to it. Many of the smaller DNA companies will disappear and others will merge with more successful members of the club. Some DNA companies will be taken over by pharmaceutical or other multinational firms. In general, large enterprises in drugs and biologicals, the food industry, agriculture, the extractive industries and the chemical industry will greatly expand their own capabilities in molecular and cell biology and genetics, and in time this total research base will exceed the specialty companies in size, though probably not in quality.

Government initiatives will ensure considerable international competitiveness in biotechnology, and a rapid overall growth of the industry. The demarcation between genetic engineering and other biologically-based process technologies will steadily diminish. Genetic engineering will come to be seen as one of many possible options for problem solving. As this happens, the almost mystical aura that now surrounds the field will fade. The hype will diminish, the special government programmes will be subsumed into the normal mechanisms for industry incentives, and market forces will assert themselves, ensuring the success of the best lines of research and the failure of many others.

Eventually, we shall be grateful for the present phase of the DNA industry. Though it will pass into history characterized by many exaggerations, it has been one force responsible for drawing increased attention to the immense power of biological science, now that the gene is understood. In the end, we shall be more glad than sorry that the academic sector had to confront mammon in this special new way. The contract between society and its institutions of higher learning has never been an easy one, and shared practical goals are one important way of bridging the gulf. We can look forward to the day when scientists will move easily between academia and industry—and back again—enlarging the contacts, softening the hard edges of misunderstanding and mistrust. Such untidy pluralism will allow the best brains to be challenged and stretched to the greatest extent, for the betterment of humanity.

11

Scientists playing God

How far should scientists go in exploring the secrets of life? Who should decide what is an ethical and safe experiment? What concerns should influence a decision to move from laboratory bench to commercial application or clinical practice? Above all, how will the awesome power to manipulate the very fabric of life affect humanity's perception of the universe and our place in it? These are just a few of the ethical, moral and philosophical issues arising from genetic engineering. None of them are entirely new, but the intensity with which they are being raised and the widespread nature of the debate exceed anything witnessed previously for the biological sciences. The picture is reminiscent of the agony of the atomic scientists fifty years ago.

There are some similarities, but also some striking differences. Nuclear physics was born in a quiet, protected academic environment and grew up in the secrecy imposed by a world war. Genetic engineering has been the subject of intense societal scrutiny from the time that it was no more than a boffin's pipedream. Moreover, it was the scientists themselves who sounded the alarm, and the history of the early worries is worth reviewing.

A unique voluntary moratorium

The biochemists chiefly responsible for forging the molecular biology revolution spring from a different background and possess different skills from the traditional medical microbiologists who isolate disease organisms and learn how to handle these safely. Accordingly, it was not surprising that when animal viruses and mammalian-cultured cells became popular tools, the molecular biologists began to worry about laboratory safety. In fact, in January 1973, a meeting was held in Asilomar, California, to debate the biohazards of certain kinds of virus work, but curiously neither this conference nor the book published from it occasioned media comment or other lay interest. By June of the same year, things had moved faster than anyone had expected. The extraordinary power of the restriction endonucleases (Chapter 3) which permit precise excision of genes came into full focus, and the first recombined DNA molecules became a reality. The relevant results were discussed at a Gordon Conference in the United States, and the participants at this meeting voted to express publicly their concern about potential risks. This was done by way of a letter sent to the Presidents of the United States National Academy of Sciences and National Institute of Medicine which was published in the widely-read magazine, *Science*, in September 1973.

This initial letter was phrased with great care; its arguments have stood the test of the intervening time. It describes the technical ability to join together DNA molecules from diverse sources, and gives, as an example, the fusion of viral and bacterial DNA. It urges the Academies to establish a committee to study possible hazards, which, however, are firmly flagged as conjectural. Despite major articles in *Science* and *The New Scientist*, public reaction was still muted. An Academy committee was set up to examine potential hazards of genetic engineering. It was under the chairmanship of Dr Paul Berg, one of the pioneers of gene splicing, who was later to receive a Nobel Prize.

This committee published some of its key recommendations in *Science* in July 1974. This has come to be referred to as the 'Berg letter'. By that time, both phages and plasmids were working as vectors for recombined genes. Some scientists were deeply worried that the commonest host bacterium was *E. coli*, a resident of the human intestine. What if a recombined organism, carrying, for example, genes from a cancer-causing virus, were to escape from the laboratory and spread like a plague? Could an epidemic of cancer result? Faced with such awesome hypotheses, the Berg committee called for a voluntary world moratorium on certain defined experiments deemed to be hazardous. They also recommended the establishment of a recombinant DNA monitoring committee within the United States National Institutes of Health, and foreshadowed a representative international meeting in 1975 to review the situation.

The Berg letter is historic, not only because of the gravity of its subject-matter, but also because it represents the first time that a senior, responsible group of biologists has called for the voluntary deferral of experiments, some of which were of extraordinary potential interest. Furthermore, until that time, concerns had been kept largely within 'the club', but now the world at large was invited to share it. Now that the cat was out of the bag, press comment accelerated markedly, particularly in New York and Washington. As far as can be determined, the moratorium was obeyed within the United States.

Most of the early genetic engineering technology had been developed in the United States, but the clever tricks were easy to copy so the reaction of European scientists was watched with great interest. They had no fiscal reason to follow a United States-based moratorium. Whereas a researcher breaking the rules in the United States faced curtailment of grant funding, moral suasion across the Atlantic was the only way Europe could be influenced. Predictably, reaction to the Berg letter in Europe was distinctly mixed. Some leaders obeyed the moratorium; a few others kept working away without making too much noise about it. In the event, the United Kingdom took the lead under the chairmanship of Lord Ashby and set up a high-powered working party, which published some eminently sensible and not unduly restrictive recommendations in January 1975. Nevertheless, it is probable that the moratorium in the United States, which lasted from July 1974 to February 1975, allowed European molecular biologists to catch up at least a little in a field which was moving with dramatic speed.

Asilomar was, once again, chosen as the site of the major international forum at which the whole world could debate genetic engineering. The world of biology really proved its solidarity at Asilomar. Most of the developed countries, including the Soviet Union, sent their top scientists; and both the press and the legal profession were amply represented. After six days of scientific presentations and vigorous discussion, a consensus document was approved with only a very few dissenting votes. The conference decided to lift the moratorium and to provide a stringent set of guidelines so that experiments could be pursued safely. It recommended two types of containment of recombined organisms: biological containment, which means disabling the host bacteria or the recombinant vectors in such a way as to make it impossible for them to survive in anything other than an artificial laboratory environment; and physical containment preventing the exit of microbes from the laboratory where the research was being performed. The types of containment recommended varied among experiments considered of minimal risk, low risk, moderate risk or high risk. Special laboratories designated P1, P2, P3 and P4 to reflect progressively more elaborate containment were described. Roughly speaking, P1 facilities represented stringent, common-sense cleanliness; P2 laboratories had features in common with a surgical operating theatre; P3 containment resembled a giant germ-free isolator and P4 units outdid the most stringent biological warfare establishments. Finally, certain experiments involving highly pathogenic microbes were banned altogether, despite the overall lifting of the moratorium. Later actions by the Director of the National Institutes of Health set up a detailed code of laboratory practice, along the general lines recommended by Asilomar, but buttressed by much greater detail, and significantly stricter.

Lifting the moratorium unleashes opposing views

Perhaps predictably, the scientists who supported the original Berg letter did not receive much praise from society for their openness and caution; but the much larger group that lifted the moratorium certainly faced bitter opposition from lay groups and from a small but significant number of leading scientists. Despite the eloquent pleas of authorities like Nobel laureate Joshua Lederberg, who referred to the certain and enormous promise of genetic engineering, and the entirely conjectural nature of the risks, the prophets of doom received the lion's share of the publicity. Noted microbiologist Robert Sinsheimer urged a continuation of the pause, because of what he referred to as the potential for biological havoc that novel forms of rapidly-dividing organisms could wreak. He also raised profound sociological questions: 'How far will we want to develop genetic engineering? Do we want to assume the basic responsibility for life on this planet—to develop new living forms for our own purpose? Shall we take into our hands our own future evolution?' Erwin Chargaff, an early pioneer in DNA research, urged a delay in further research, if necessary of two or three years, till some host other than *E. coli*, perhaps a marine organism, could be found.

The most direct confrontation between scientists and the lay public occurred in 1976, in that Mecca of pure science, Cambridge, Massachusetts, the home of both Harvard University and the Massachusetts Institute of Technology. The trigger point was a proposal to build a P3 laboratory. Opponents of recombinant DNA research, including some very prominent scientists, raised the matter with the city council, which took the unprecedented step of imposing its own three-month moratorium on genetic engineering, a move apparently seen as legal within Harvard and MIT. The end result was a local review board that set its own safety and health standards, over and above those of the National Institutes of Health.

The idea of local councils buying into the debate did not spread. The United States and United Kingdom examples of containment guidelines enforced largely by peer group pressure (and grant cuts for transgressions) were taken up in many countries, including Australia. On the whole, this system appears to have worked well, with only one or two examples of failure to comply having come to notice. Detailed and far-reaching legislation was, indeed, seriously considered in the United States, but not proceeded with. This was also the case in several European countries.

The genie turns out to be benign

In the event, none of the fears of the prophets of doom materialized. The capacity of *E. coli* K12, the strain most commonly used as a host for recombined vectors, to establish itself in the human gut was shown to be quite limited. Individuals fed massive doses of this strain excreted a few viable organisms in their faeces for a few days, then the E. coli disappeared. In fact, this is not surprising, because the whole history of microbiology suggests that when bacteria are grown repeatedly and for long periods on artificial media, they progressively lose the capacity to infect the original host. Furthermore, microbiological containment really does work and even were it imperfect, there is an immense difference between swallowing a billion *E. coli* intentionally, and ingesting one or ten organisms accidentally. The experience of literally thousands of laboratories now engaged in recombinant DNA research has proved that the technology as such is entirely safe, and not a single health incident has been reported since the moratorium was lifted. Of course, the spectre of someone using the technology for evil purposes, such as germ warfare, cannot be discounted. All one can say is that no evidence of this type of research has surfaced. With the benefit of hindsight, it is now possible to state that none of the conjectural hazards materialized, and that the initial containment guidelines were unnecessarily stringent. Indeed, in early 1979, the National Institutes of Health relaxed the levels of containment somewhat, many experiments edging down one level, e.g. from requirement for a P3 laboratory to P2, etc. Most other countries followed suit. We confidently expect this process to continue, and believe that the great majority of recombinant DNA work will eventually require little more than commonsense cleanliness in the laboratory, that is, P1 facilities. As a matter of fact, recombinant DNA research is now such a commonplace component of virtually every medical and biological research field that it must be difficult for younger scientists to understand what all the fuss was about a quarter of a century ago.

Another success story has been the apparent worldwide voluntary compliance of industry with the guidelines in each country. This is admirable, as submitting research protocols to monitoring committees must have occasionally involved revelations of corporate strategy normally kept secret for commercial reasons. Naturally the manner in which gene research is regulated today will vary from country to country. For example in Australia, the nation has been very well served by a Genetic Manipulation Advisory Committee (GMAC) which adjudicates on proposals put before it and which commands wide respect although its advice lacks the force of law. Concerns over GMOs and a desire to oversee their testing and introduction closely has led to the establishment of a more high-powered body, the Office of the Gene Technology Regulator, to function under Federal jurisdiction but with major input from the States. This body will more formally regulate both applied research and the more basic aspects of genetic engineering. This success story in self-regulation should not make us complacent. It does not mean that all uses of genetic engineering are without risk and that we can safely apply it in any circumstances. There may be techniques such as the wholesale alteration of the human organism that cannot ever be accepted. However, these decisions will be made and remade by society and practices that are not now acceptable may later be adopted. This has ever been the way of the world.

Opposition to genetic engineering in a historical setting

We will now return to the questions posed at the beginning of this chapter. It is a fact that this technology, and extensions of it which can be logically foreseen, give humanity the capacity to find out more about the basic processes of life than ever before, and to create life forms in ways that nature never intended. These vast new powers frighten many people.

In fact, there is nothing new about a distrust of science. In his brilliant essay, 'Reflections on the Neo-Romantic Critique of Science', Leo Marx reminds us that many of the eighteenth-century writers questioned 'the legitimacy of science both as a mode of cognition and as a social institution'. Alfred North White-head saw this romantic reaction as 'a protest against the exclusion of value' from the sober array of facts which are the fruits of scientific work. Somehow, the Arcadian vision of nature as good, supreme, bountiful, was challenged by the impersonal machines of technology that were eroding nature's grip on human destiny. While the reliability of scientific knowledge within its own frame of reference has really not been seriously questioned since the days of Galileo, Whitehead sees a discrepancy between what science provides in the way of knowledge, and what mankind actually wants by way of a meaningful existence. Strangely, the romantic writers make scant mention of the effects of science and technology on the crushing burdens of work which the poor were forced to assume in order to eke out a meagre existence. More recently, C. P. Snow lamented the gap between the 'two cultures', but even he did not foresee the intensity of the dissident movement, the intelligent anti-science counterculture, which reached its crescendo in the Vietnam war era. This saw science as the villain responsible not only for the tools of destruction but for fostering a mentality that could allow them to be used.

Basically, the opposition to major technological change, and thus indirectly to science, comes in two forms. One the one hand, there is a tendency for people to fear the unknown, to resist change, to preserve comfortable preconceptions, to resent new circumstances not of their own making. This kind of objection is best countered by modulating the rate of technological change, and while this is a major challenge for politicians and other decision-makers, the body politic has the capacity at least to address the issue and make appropriate choices. The challenge becomes greater when the technological change is being driven by a model based on free-enterprise capitalism. Research and development programs within companies have the capacity to produce new products. However the costs can be high and the company has the exclusive right to market this new product for the life of its patent, currently twenty years. If the pace of progress is slowed, then the time available to market the product is decreased and the company's chance of recouping development costs and making a profit are reduced. Accordingly in technology-driven fields such as bio-technology there will be an economic push for rapid change. How can governments and regulatory agencies moderate this drive for change and still maintain an atmosphere of innovation? The answers are not easy but may involve some alteration to the patent law that can allow some sort of hiatus while the technology is assessed for societal effects beyond the area of product safety.

The second class of objection is more difficult to counter. It is more abstract, and relates to whether a scientific view of nature, after all a rather recent event in human affairs, somehow robs humanity of other ways of finding truth or knowledge. Theodore Roszak wonders whether an ingrained commitment to science as the reality principle 'frustrates our best efforts to achieve wholeness'. He links humanity's recent flirtation with science and technology to the sometimes terrifying trend to urbanization and deplores the 'technocratic elitism' which characterizes not only the industrialized countries but also dominates the leadership of most developing countries.

It would be altogether too facile to dismiss this category of opposition to science as being Luddite responses of disaffected minorities. Many able and intelligent people perceive a genuine threat. For example, Pope John Paul II, in an address to UNESCO, expressed the following view:

> The future of man and mankind is threatened, radically threat-ened, in spite of very noble intentions, by men of science . . . their discoveries have been and continue to be exploited—to the prejudice of ethical imperatives—to ends . . . of destruction and death to a degree never before attained, causing unimaginable ravages . . . This can be verified as well in the realm of genetic manipulations and biological experiments as well as in those of chemical, bacteriological or nuclear armaments.

This might appear an extreme view, but we must ask why the various countercultures, having in common a rejection of popular material and economic goals, enjoy so much support. Furthermore, a decline in participation in the 'harder' natural sciences has been noted in schools, as well as an increased interest in the 'softer' social sciences. So disillusionment with science and technology runs deep in some segments of society. Demonstrations at meetings of the World Trade Organization and the World Economic Forum brought together those worried about the pace of change and globalization of trade and culture with those worried about the dehumanizing aspects of scientific advance. Clearly many people are worried.

In defence of scientific truth

This vision of science as somehow anti-human, coldly perverting people from a truly satisfying destiny, must be refuted, because it offends common sense—it is simply not true. If fault there be, it lies in mankind's nature and the uses to which power can be put.

Science and technology have been embraced by people all over the world for one simple reason: they work. Sir Peter Medawar has argued that 'science, broadly considered, is incomparably the most successful enterprise human beings have ever engaged upon'. We do not have to go all the way with Marcelin Berthelot who declared that science 'will provide the truly human basis of morals and politics in the future'. Nevertheless, it is unfair to blame science and technology for ills in the human condition that are as old as mankind: for undue aggression, selfishness, greed and a chronic incapacity to live up to one's highest aspirations. It is as illogical to blame science and technology for not slaking our thirst for spirituality and transcendence as it would be to blame literature, art and music for not feeding, clothing and sheltering us. Science can only address part of the phenomenon of the human condition. Few people realize this more fully than the scientists themselves, who, as a group, are better read and more concerned with humanistic values than many other technical and professional groups.

There are undoubtedly some who will say that genetic engineering research offends nature, that the creation of new life forms should be left in the hands of evolution, not in mankind's. We must listen to this view, but also be careful to explain to its proponents that conventional genetic techniques employed by civilizations for ten thousand years have already had a formidable impact on the ecosystem, and it might indeed be difficult to distinguish, say a disease-resistant strain of wheat created by scientific breeding and selection pre-1975 from one fashioned tomorrow through the new technology. Should we really stop ourselves from accelerating the search for moulds which make better or cheaper antibiotics because DNA splicing is somehow intrinsically bad? Well, if it appears that this makes little sense, should we at least declare 'hands-off' genetic engineering of higher life forms such as mammals? But if genetic manipulation of growth hormones were to allow a steer to grow to full size in six months rather than three years, is it evil to create such animals given that we already have feed lots, and that the world is hungry for first class protein?

What then, about human beings as genetic guinea pigs? This is clearly the area that has caused the most concern and also confusion. We can foresee as realistic possibilities the manipulation of cells and tissues of a given sick human individual, one of whose genes is unhealthy. This seems worthwhile and noble if it can be achieved. There is currently no approach which can cure single gene defects inside a person and repair the genes in all the sperms or all the ova. Accordingly, the only way of eradicating bad genes (or repairing them) for the benefit of future generations is to contemplate treatment of sperm, ova or early embryos in the test tube, before **artificial insemination** or embryo transplantation. Given that practical ways of doing this could be decades away, and that sperm or ovum selection might be more practical in many situations involving recessive gene traits, there still appears to be nothing that humanity should fear in this approach. The issue of human cloning is one that we are going to have to face much more quickly than seemed possible a few short years ago. However we should stress that currently humans have not been cloned and there are many regulatory and legislative barriers to this task, at least in the developed world.

Somatic versus germ-line gene therapy

That being said, most scientists do make a categorical distinction between somatic gene therapy and germ-line gene therapy. Somatic gene therapy seeks to alter the genetic constitution of cells of a particular sick individual, inserting a 'good' gene for the one that is faulty. The effects, for good or for harm, are confined to the individual being treated. Somatic gene therapy is considered ethical by most scientists. Germ-line gene therapy seeks to repair a gene defect at the earliest stages of embryonic development. The effects, good or bad, affect both the child which develops and also all future generations. Most scientists oppose germ-line gene therapy, first because an accidental ill-effect would be passed on indefinitely, and secondly because we do not yet know enough about functional genomics to predict all the unexpected side effects which could occur. It does appear, however, that this near-absolute opposition is softening somewhat. However, there are still very few proponents of positive eugenics based on gene therapy.

Now we come to the 'thin end of the wedge' style of argument. 'If you are curing thalassaemia today, will you not be tackling social rebelliousness tomorrow? If we allow this kind of thing to start, where will it all stop?' The first defence, but not the most important one, we believe, is that we still have only the sketchiest of notions about the processes of inheritance that govern complex features of character, or even most physical attributes such as strength, beauty, tendency to obesity and so forth. These are clearly the results of the interplay of a multiplicity of genes, and of societal and environmental forces impinging on each individual. They are therefore simply not amenable to gene therapy, and might never be. Even were this not the case, we find it difficult to support a line of reasoning which says: 'Do not do this good thing, because it might lead you on to do that bad thing'. The whole history of humanity has been to probe, to examine, to explore, to seek the limits of understanding and then to exceed them, each generation building on the legacy of its predecessors. To deny that thirst for knowledge is to destroy humanity's wholeness, more surely than anything else.

And what if the chemical nature of humans is the object under study? Is it healthy for us to know we are 'just' a few DNA molecules being copied and read? The answer again is blindingly clear. Of course it is good for us to know more about what we are: at worst, this knowledge might allow us to prevent and cure our most obvious ailments; at best, it might even help us to deal more effectively with one another. No knowledge of a natural truth gained by objective search can be harmful, though its misuse obviously can. Furthermore, no depth of insight into the physical nature of ourselves that we can derive from scientific experimentation will detract from or complete with the insights that we gain through the humanities, though indeed a complementation is an eventual possibility.

Just as the decision on whether to pursue recombinant DNA research was never really in doubt, but only an uncertainty about timing and order, so the question of whether to apply the technique to humans is not in doubt either. It does raise a whole untidy series of issues which must be faced, one by one, and which will surely look rather different when they are no longer hypothetical. How, then, are we to ensure that the genetic engineering revolution is kept on track, as the servant of society; is harnessed towards noble ends; and is developed at a pace that our civilization cannot only bear but actively welcome? These issues of public policy are discussed in the next chapter.

12
Genetic engineering and public policy

Scientists occupy a strange, even difficult, position in our society. On the one hand, they enjoy considerable freedom to pursue their own interests, exercising their creativity unshackled by the daily demands which constrain most workers. On the other hand they, or at least those in the universities and research institutes, are sustained by public funds, so must in the broadest sense be responsive to society's needs. And, most assuredly, scientists are not above the law. Without a doubt, the power of the scientist to dictate humanity's agenda has grown formidably. Do we now need new laws and new public policies to control the scientists?

Existing levels of control on scientific experimentation

It is important to point out that, quite properly, today's research worker already faces a battery of controls, acting at many levels, some direct and some indirect, which bear on his or her activities on a daily basis. It is worth examining these in some detail. First, each organization performing research has safety rules, a professional safety officer and a safety committee. All research in an institution must conform to standards imposed by these internal elements. It is now frequent to encounter biohazards committees distinct from the traditional safety committees used to worrying more about fire, explosions or toxic chemicals. Secondly, all research has some kind of funding source, and if large governmental agencies are involved, the granting body will usually have an ethics committee and various more specialized committees certifying the safety of given categories of work. This is the chief lever that has been used so far in recombinant DNA research. Thirdly, if research on human subjects is contemplated, it is necessary to obtain both the informed consent of the person, and the approval of an ethics committee of the hospital or university, such committees always having lay members as well as medical specialists. Fourthly, scientists are answerable to the normal administrative controls of institutions and thus subject to discipline if they embark on a course of action considered outside accepted norms.

More indirect controls have equal force. Peer group pressure is ever-present in science, peer group approval one of the most eagerly sought rewards. Now that areas such as genetic engineering, foetal research and organ transplantation have been brought so much into the spotlight, the peer group itself, thoroughly indoctrinated with locally accepted guidelines, will be a monitor of any would-be dissenter. The publication process, also, is an indirect level of control. For example, experiments involving any unnecessary cruelty to animals may be rejected by an editor, as would any manuscript failing to detail safety precautions in potentially risky experiments. The fiscal mechanism also contributes in an indirect way. If a line of work, though not directly contravening a safety rule, nevertheless seems near the borderline, its originator may well find the going a little tougher when final priority score for funding is assigned.

All the above controls arise, to a greater or lesser degree, from within the scientific enterprise. What of controls from the outside? Again, in some areas such as permissible levels of radiation or toxic wastes, there are specific regulations and laws imposed at federal, state or local levels. Genetic engineering has so far largely avoided this level of control. However, there is also the precious heritage of the common law. If a recombinant DNA research group were to flout the safety standards, regulations and guidelines accepted in a particular country, and some harm were to come to one or more individuals, the injured party could sue, and would probably win. One of the surest proofs of the safety of genetic engineering research carried on to date is that this has not yet happened!

Specific legislation or flexible guidelines?

This set of answers does not satisfy those critics of science who believe that the public should have more say in what scientists do, or are allowed to do. The critics argue for specific legislation covering areas deemed controversial or risky. Such legislation has been in force for years in the field of organ transplantation, and appears to have been well accepted by both patients and doctors. Legislation also covers foetal research and areas such as artificial insemination, *in vitro* fertilization and induced abortion in most countries. This has resulted in a great deal of controversy, bitterness and even hatred. In some instances, widespread flouting of the law has occurred. Are new laws the answer to better development of genetic engineering?

Proponents of specific legislation make a number of cogent points for their case. Microbiological agents such as disease-causing organisms are clearly dangerous and are already covered by various laws and regulations. Should not the new potential for creating still more dangerous variants be covered by extensions or amendments to these laws? Moreover, if technology has the potential to cause catastrophic damage (eg release into the environment of an extraordinarily toxic engineered microbe; or uncontrolled weed-like spread of some plant variety with superior growth characteristics), might a point be reached where society should seek protection under the law rather than relying on voluntary restraints? Laws might also ensure that monitoring committees would include a full spectrum of community views, and would have real powers. Furthermore, an indirect argument holds that many of the implications of rapid scientific development are not being addressed sufficiently by lawmakers; perhaps the dramatic example of genetic engineering could serve as a triggering point to alert the legal profession and the politicians to their responsibilities. Finally, a call for new legislation might be one factor encouraging greater public participation in the debate about where science is taking the world. All these views require careful examination.

Having followed this debate closely for twenty-five years, and having developed great respect for some of the able proponents of an opposite viewpoint, our conclusion is against the introduction of new laws on genetic engineering. There are many complexities and no advantages in going down the legislative route in this field. First, the technology is changing so rapidly that legislation would be extremely difficult to draft: the boundaries of genetic engineering are constantly being extended and the distinction between it and the rest of cellular and molecular biology are becoming blurred. For example, *infusing* DNA from one cell into another would be classified as genetic engineering, but *fusing* the two living cells to achieve genetic hybridization would not. Legislation would probably require amendment before the printer's ink was dry. Secondly, society's worst nightmares know no geographic boundaries, but laws do. If a particular nation were to enact restrictive legislation in the recombinant DNA field today, this would not stop the research; it would simply move it to another country. There are instances of this happening; for example, where one country demanded P4 level of containment, the most rigorous and expensive laboratory facility, for an experiment which another classified as P3, a substantially less elaborate set-up. The only end-result of a given country taking a 'tough line' with its scientists would be that the country would simply hand the lead in the forbidden area to countries with a different viewpoint. Thirdly, the passage of legislation would still not offer the critics what they want, namely, control of issues by non-scientists. Much of the legislation would have to be highly technical, and provisional until conjectural hazards were subjected to more testing, so that experts in the field would essentially have to do all the drafting and make all the tough decisions. This comes back more or less to the present position! Fourthly, the horror situations that have been painted lack realism. It is very doubtful whether it would be possible to tailor-make plant varieties that truly spread like wildfire, or to invent germs more horrible than the ones like Lassa fever, anthrax or botulism, that we already possess. Finally, it is important to recall that none of the conjectural hazards that have been mooted have in fact materialized. Conjectural hazards are one thing but human ingenuity turned to evil purposes is another. Experts on defence against biological warfare in the United States have re-ceived intelligence that some rogue states have stockpiled huge quantities of smallpox virus and anthrax bacteria. Given the global eradication of naturally-occurring smallpox and the consequent cessation of vaccination against smallpox, virtually no one under twenty years of age has any protective antibodies, so a foreign power could immunize its own population and spread the virus to the enemy population. Similarly, with anthrax. The United States is taking this threat sufficiently seriously to stockpile many millions of doses of the vaccinia vaccine. The point of this story is that we do not need genetic engineering to create hazards of truly horrendous proportions.

Rather than legislation, we would prefer the soft-edged, untidy, polyvalent methods of a free and decent society. There is a fundamental residue of idealism within the biomedical research community, and the threshold of consciousness about risks has been raised to the extent that most silly things will be stopped before they happen. Both scientists and the regulators have gained much from over a decade's intensive debate and the fundamental challenge now is not to control but to promote recombinant DNA research.

The commercialization of science

We have several times throughout the book discussed the issues related to commercialization. As mentioned earlier (Chapter 10) there have been an enormous number of new biotechnology companies founded, some becoming hugely successful and others becoming bankrupt. The industry is fast moving and success depends on a good idea, the capital to pursue it to fruition and a dash of luck. Many of the mechanisms we have outlined for the control of the progress of science simply do not apply to these biotechnology companies. Although almost all use a system of monitoring occupational health and safety and ethical issues, there is no absolute requirement for them to do so. Their adherence to standards is voluntary. There is no requirement for them to expose their experiments to the scrutiny of their peers by publication. In fact, open disclosure of their work in full detail almost never occurs. Their procedures and results represent their intellectual capital and this is rigorously protected.

Funding agency prohibitions on the performance of certain lines of experimentation such as human cloning or stem cell work also do not carry any real weight, as these companies do not often rely on such sources of funds. Indeed, it could be argued that for a few companies work in such areas would be attractive. There would be few competitors and it would be possible to steal a march on the opposition, in the expectation that eventually the area would be opened up.

There do remain the areas of government legislation and the common law to ensure proper behaviour among companies. One could take a pessimistic view of events: that companies potentially have the capacity to generate new organisms where the capacity for harm could be far greater than the ability of the company, particularly a small start-up, to make recompense. This might be so, but it remains a highly speculative possibility. We would emphasize rather that as far as can be told, the vast majority of companies act responsibly and follow current conventions on what fields of inquiry are appropriate. There is no evidence of widespread offering of 'designer' babies or cloning of humans.

The recent behaviour of the Monsanto Corporation serves as a salutary example of responsiveness to community concerns. This company has developed a number of new plants with useful properties that decrease farming costs (Chapter 9). They wished to protect this investment and were concerned that people could obtain seeds from growing crops and use them to plant the next year's crops. Thus after one growing season, no-one would have the need to buy further amounts of the product. Accordingly, the company began to develop a technology that would render its plants, both conventional and transgenic, sterile. Those wishing to grow plants for the next harvest would need to buy fresh seeds from Monsanto. This technology, dubbed the 'Terminator', was highly controversial, as it acted to end practices as old as recorded agriculture. The company appeared adamant that it would deploy this technology within its products. However after several years of vigorous debate, the company decided that the negative publicity was more damaging than any profits it might gain. Accordingly it abandoned development of this technology and pledged never to use it in any future products. Thus, peaceful protest, selective boycotts and moral suasion were able to influence a huge multinational corporation. Such efforts are likely to be even more effective on small companies that might cross the lines of probity. Parenthetically, it is worth noting some of the complexities and oddities of this debate. In another arena, the opponents of GMOs complain that there is no way to stop cross-pollination of transgenic and non-transgenic plants so that adjacent farms may be 'contaminated' by pollen coming from plots of transgenic plants. The solution to such a worry is of course technology just like 'Terminator', that renders the transgenic plants sterile. So one technique is both reviled and desired by critics of GMOs, depending on the circumstances.

We have commented in Chapter 11 about the pace of technological change being driven by a company's desire to both recoup development costs and make a profit within the period of patent protection. We would note one other aspect of the profit motive. We have spoken about the potential of the genetic engineering revolution to affect human health and wellbeing. Arguably, it is among the populations of the developing world that the greatest strides in health and improvement of nutritional and economic status could be made. These are the poorest, hungriest and sickest parts of our globe. Yet it is unlikely that many companies will focus on developing products for these people, because they do not have the capacity to pay high enough prices to ensure company profits. The failure of pharmaceutical companies to develop new drugs and vaccines for diseases of the Third World is well known and lamentable. It appears that the biotechnology industry will continue this same pattern of behaviour. The one potentially bright spot is that the costs of engaging in biotechnology are not high and it is not beyond the resources of even the poorest country to begin to address their own problems in health, agriculture and environmental protection. It is critical to ensure that enough enabling technology remains in the public domain so that these countries are not hamstrung in their efforts. Overly inclusive patents can be a problem in this regard. We note the granting of a European patent to Agracetus Inc., a wholly-owned subsidiary of W. R. Grace Co. of Boca Raton, Florida. This patent, number 301 749 B1, covers all forms of genetically engineered soy bean plants and seeds—irrespective of the genes used or the transformation technique employed. This extraordinary patent essentially grants a single corporation control over genetic research on one of the world's most important food crops. This would seem to be an unacceptable extension of the patent system. Theoretically, another country could develop a modified soy bean and plant it for crops, as long as they did not export the products. However, they might find themselves subject to boycotts and embargoes aimed at preserving patent rights. A coalition of groups has now lodged an appeal against the granting of this patent, and hopefully sanity will prevail. It does demonstrate, though, that large areas of important research for developing countries might be tied up by commercial concerns with no interest in supplying those countries with goods they can afford. It should be noted that such problems of adequate access to technology for Third World countries are not new, but it is disheartening to see them continue into this millennium.

Biological species as intellectual property

Biotechnology companies operate in an arena where often their patents and their intellectual property are their only value. It is widely held that without patent protection, such a company has no real value. Certainly in any negotiations with other companies, the strength of the patent position is directly related to the strength of one's bargaining position. However this plethora of companies and patents can also have a detrimental effect on innovation. When the patents required for development of a new product are spread between many different companies, it is often hard to reach a suitable licencing agreement. Companies can overvalue the worth of their patents and ask too much for their use. This proliferation of fragmented and overlapping intellectual property rights can stand in the way of new product development as there are often high transaction costs in bargaining among many owners. In these cases commercialization can stand in the way of achieving the public goals of biomedical research, which is the betterment of humanity's condition. It is important for policy-makers to observe this state of affairs carefully and act if restrictive licencing practices start to interfere with product development. Otherwise, more companies and more patents may lead paradoxically to fewer useful products for improving human health.

The other effect of the importance of patent rights is a highly litigious atmosphere between the developers of new products that often might rely on prior art. The plaintiffs and defendants include both companies and academic institutions. A perusal of recent court cases in the United States reveals a plethora of litigation about patent rights and infringements. A non-exhaustive list includes the following litigants: Amgen v. Chugai Pharmaceutical Co. over erythropoietin; Scripps Clinic & Research Foundation v. Genentech over Factor VIII; Genentech v. Wellcome Foundation over tissue plasminogen activator; Hormone Research Foundation v. Genentech over human growth hormone; Genentech v. Eli Lilly

& Co. over human growth hormone; Bio-Technology General v. Genentech over human growth hormone; Genentech v. Novo Nordisk over human growth hormone, and Genentech v. Chiron over insulin-like growth factor. One might conclude that a career in patent law is equally as promising as a career in biotechnology.

Before leaving the legal arena, it is worth noting that in 1980, by the slimmest possible majority of five to four, the Supreme Court of the United States decreed that a living organism could constitute a patentable entity. This opened the way for commercial enterprises to increase their involvement in genetic engineering under the same kind of patent protection as applies, for example, to drugs or chemicals. It also legitimized a patent that Stanley Cohen and Herbert Boyer had taken out covering the broad technology used in most genetic engineering work. This meant that Stanford University and the University of California at San Francisco would get licence fees and royalties from essentially all products of genetic engineering that reached the market for a period of twenty years.

This view of the patentability of living organisms did not extend to Europe, at least initially, and it was only somewhat later that the European patent office began granting patents for living organisms. Currently there is still not total harmonization between the two major patent systems in the world and some things patentable in the USA are not patentable in Europe. The patenting of living organisms raises some tough legal issues, far beyond the scope of a book such as the present one. The interested reader is referred to a number of sources outlined on the Reshaping Life web site including a fine work sponsored by the Cold Spring Harbour Laboratory entitled *Patenting of Life Forms*, edited by David W. Plant and others. Although somewhat out of date now, it collects the views of both interested scientists and some of the most distinguished jurists working in the field and stands as a thought-provoking study of the area.

The need for increased communication between science, industry and government

In Chapter 10 we noted that various governments are actively promoting components of the DNA industry in their countries. In view of the storm that surrounded the birth of genetic engineering, and the enduring concern for more regulations and legislation in some quarters, should more be done to keep politicians and jurists abreast of this field? What can we do as a society to make decision-makers at many levels of society more aware of this and other issues involving high technology and scientific specialization?

This issue, we believe, is currently far more important than the one of new legislation. To address it, we must first ask why a need for more extensive communication and debate exists. The central difficulty is the ignorance—and indeed apathy—within the community about the research process and about the nexus between fundamental research and applied development. Our lives are affected by technological change at a daily level and in the most diverse ways. The future will bring further technological changes. Some of these we hope for fervently (a cure for cancer; pollution-free vehicles) and others we anticipate with a mixture of admiration and apprehension (more robots in factories; rocket-powered travel). In either case, most of us are concerned exclusively with the practical end-result of scientific development, with the completed experiments that impinge directly on our lives. But all these changes, from the most important to the most trivial, depend on people: on scientists and technologists staffing the formidable enterprise of global research and development. Almost by definition, these people are specialists. They live within a linked spectrum of subcultures, and, in general, exercise their influence within the world of ideas rather than through the acknowledged realms of power, namely politics or business. Thus, there are very few scientists in the world's parliaments or, with some exceptions, in leadership positions within public service. Perhaps more surprisingly, there are few scientists in management positions within industry or commerce, even in high technology industries. So the people with the technical capacity to shape the future are under-represented within the highest levels of decision-making.

Scientists and technologists know full well the dynamic interplay between curiosity-motivated basic probing of concepts and mission-oriented study of some hoped-for practical advance. In general, they know what can be anticipated with near-certainty provided enough effort is expended; what represents a fair gamble for the medium term; and what is still beyond reach. They know further that the only hope of bringing the last category of advances to eventual fruition is by sustaining a broadly-based, patient effort in fundamental research aimed simply at understanding nature in all its manifold manifestations. But the general public knows little about this, and, worse still, cares little.

This separation of the process of research and technological change from the process of leadership and governance of society imposes the need for better communication. It should be a communication between equals, pursued with objectivity and neutrality. The scientists should inform the public about what they are doing; about what practical results can realistically be expected; and about what longer-term dreams can legitimately be espoused. Society, through all the means of communication and leverage available, should respond and thereby guide scientific enterprise into those broad subject areas requiring most emphasis. This process is, of course, going on in most countries, but labours under certain constraints.

Communication between science and industry is also important. Fortunately, a new, government-sponsored research initiative has made a significant impact on this in Australia. It is the Cooperative Research Centre or CRC scheme. There are over sixty CRCs in Australia, each specializing in an important branch of science or technology. The idea is to have intense collaboration between university/academic science, applied science in government laboratories (such as CSIRO) and industry. Perforce the participants in CRCs have had to learn one another's language. The PhD students in the CRCs are of particular importance. They emerge from their training steeped in the norms of both science and business. Over time, these young people should make a major contribution to bridging the gap between theory and application.

Structural problems impeding objective communication

The first constraint relates to the knowledge base within society. Most people, including many influential leaders, lack the patience and the intellectual stamina to follow scientific arguments, even if these are structured in simple, non-technical language. In many respects, though so deeply influenced by the end-results of science, we are a scientifically illiterate society. As people who have struggled hard to make modern biology accessible to lay audiences of diverse types, the authors are only too familiar with the glazed look or patent discomfort that is the frequent reward for our efforts! Some mental trapdoor seems to evoke memories of schoolroom tortures, defences spring up which say: 'this is intended for someone really clever; I'd better leave it all to the experts'. The simplest response for the scientist is to get to the 'bottom line'—the practical implications of the research—very quickly, thereby losing much of the story.

The scientist, too, bears much of the blame for suboptimal communication. Two deficiencies are particularly common. First, the scientist either will not or cannot make the adjustments required to alter the mode of communication from a technical to a popular one. One of the cardinal sins in a scientific paper is to make exaggerated claims for the generality of a particular discovery, or to omit the limitations of the scientific approach used, which may require qualification of the general conclusion. Therefore, scientific papers use a contrived, stylized, rather coy prose form full of phrases like: 'Table III appears to indicate that . . .'; 'the results are consistent with the following interpretation . . .'; 'within the limitations of the present experimental design . . .'. Furthermore, the paper will usually be full of technical terms that seem quite commonplace to the investigator, so much so that their obscurity to the lay person occasions surprise. When discoveries are translated for popular consumption, it is better to go to the heart of the matter and omit the minor qualifications, provided that the listener or reader is not left with an inflated idea of what has happened. It is also necessary to use familiar terms and perhaps homely analogies to get the message across. Secondly, scientists find it hard to be objective and neutral about their own work, and often foray into the realms of communication with the lay person only when they want something—usually a larger grant! This, obviously, creates a predictable reaction. Decision-makers discount the scientist's claim, sometimes appropriately, but frequently to a more than necessary extent. Too much of the communication takes place in a setting that has at least a component of an adversarial relationship; too little in the context of free brainstorming out of sheer interest.

These constraints show up most clearly in the area of pure or fundamental research. It seems to us that an innate scepticism about its value limits the commitment which politicians are willing to make to universities and similar institutions. Again, this is not all the fault of the politicians. The message that giant strides in humanity's capacity to control nature result from the most unlikely, curiosity-motivated research cannot be repeated too frequently. Who would have predicted that research on some obscure bacterial enzyme which can chop up DNA into little bits would sponsor a new industrial revolution? Analysis of all the major developments in technology soon reveal their dependence on prior basic research, and revelation of this nexus should form part of every attempt at popularization of science.

The role of the public sector in promoting genetic engineering

The above discussion is critical in tone, but at the same time the situation is improving, and, for the biological sciences, the DNA story has done much to promote greater efforts at discourse between scientists and the rest of society. The above problems do not lend themselves to easy or definitive solution. They require gradual attitudinal changes and steady effort. Indeed, this book itself is one small example of what can be done to break down the barrier in communication, an effort made easier by society's lively interest in anything that affects health. It may be a presumptuous conclusion, but it appears that society is now ready for the public sector to take an active role in the promotion of genetic engineering, in part because of the serious work that has been done from both sides of the fence.

There are three levels at which public policy can have a major effect. Each has been creatively attacked through recent initiatives in Australia, but the principles involved are general ones applicable to all countries. The first level involves the creation of a cadre of scientists and technologists capable of participating in the DNA industry at a high level of expertise. While excellent undergraduate courses are obviously necessary, they are not sufficient. The most cost-effective way of enlarging the pool of trained workers is to encourage an expansion of the research teams of established leaders in the field. This should be through the research grant mechanism of peer group review, but one with guidelines that make it clear that projects to be sponsored should have a substantial component of applied science. Such a mechanism has a built-in quality control device, and also a relatively short lag period, giving it advantages over more elaborately structured and expensive national educational plans.

The second level involves the encouragement of venture capital formation, preferably in a manner that makes large as well as small enterprises look closely at the risk-reward equation in this field. Tax incentives represent one simple and effective way of doing this; long term, low interest loans another; and equity participation by governmental bodies a third. In smaller countries, access to capital alone may not be sufficient to establish a biotechnology industry. Mechanisms that help inventors or entrepreneurs gain access to managerial expertise might be important. To help small firms develop discoveries, government departments should—and have begun to—provide advice on markets, feasibility, international trade potential, and so forth. Export incentives will be especially valuable for countries with a limited domestic consumer market. A supportive stance by government in the early years is probably necessary in most countries if they are to have any chance of competing with the United States in this field.

A third area to which attention must be given in the relationship between academic departments advancing DNA research and industrial concerns anxious to exploit these advances. This remark may seem trite, but it is curious how substantial a gap still separates these two universes of endeavour, despite the rapprochement mentioned in Chapter 10. In addition to the spontaneous developments mentioned there, specific government policies can narrow the gap by mechanisms which tempt the academic to seek out a commercial partner more actively. For example, a scheme recently introduced in Australia gives grants preferentially to academic departments which have formed formal linkages with firms willing and able to take the resulting discoveries to the stage of commercialization. Schemes which promote interchange of workers between universities, institutes and industry also have appeal. These could encourage younger scientists to cross the barriers for seconded periods of two to three years, and could provide for sabbatical periods of six to twelve months for senior workers. It would be important to have traffic in both directions and to safeguard the career prospects of the participants. Such schemes cost little, but create a climate where the traditional distrust would be broken down quite rapidly, given a society where each of the two sectors is doing its job properly.

Obviously, detailed patterns of government intervention would vary in each nation, and over time. It is pleasing to record that public policy in Australia is moving in precisely these directions. Australia is a country with a small population, and not noted for the success of its manufacturing industry. As investors, Australians are somewhat risk-averse, and the venture capital industry has been slow to develop. This does appear to be changing. The vast amounts of money controlled by the superannuation funds constitute one possible source. Changes to the capital gains tax regime are also encouraging individual investors to be more adventurous. As venture capital builds up, there will be a need for entrepreneurs who can link the generation of ideas with the sources of funds, and who can help to turn a good idea into a marketable product. There is an urgent requirement for the manufacturing sector to become more competitive in world terms. A tradition of excellence in biological research at the basic level, and a broad recognition that biotechnology offers opportunities for the creation of high value products where the distance from export markets is not such a drastic disadvantage, presage an exciting future in the decade ahead.

We should note that although legislative and regulatory mechanisms appear to be sensibly designed within Australia, the actual amounts of money devoted to supporting research and development are too low. Biotechnological research is entering the phase of industrial scale experimentation with genomics projects in overseas laboratories and companies making widespread use of robotics, computerization and advanced machinery. Almost none of this is occurring in Australia because of the cost of this equipment. Chronic underfunding of the universities is destroying the ability of this country to train skilled new researchers and many of our brightest are being lured overseas, particularly to the USA. Both public and private sectors have failed to support research and development, a state of affairs noted by many local and overseas reviews. For example, both the reports of Australia's Chief Scientist *The Chance to Change* and the Innovation Summit Implementation Group *Innovation—Unlocking the Future* have highlighted this lack. A fundamental rethink is needed about the state of Australia and its sources of wealth, both present and future. How will we maintain our standard of living and prosperity into the future? Biotechnology offers one way but it must be adequately funded at many levels. The challenge to the government and people of Australia is clear.

13

Distant horizons

Speculations about the future of genetic engineering tend to fall into three groups, each possessing able and vocal proponents. There are those most excited about the industrial and commercial potential, who see new Silicon Valleys emerging, capable of jetting whole nations into a new golden age of prosperity. In fact, the term Genome Valley is starting to be widely used! Then there are those whose dominant concern is with twin dangers. They fear a wilful or accidental release of highly pathogenic species into the biosphere. Perhaps some of the engineered GMOs made to help mankind will in some way alter the ecology of the planet and destroy our ability to grow our food. They also worry that an excessively mechanistic appreciation of the nature of life may further drive humanity towards materialism and a sterile, stereotyped view of the universe, devoid of subtlety or value. Finally there are the scientists themselves, many of whom believe that molecular biology is the new key to the solution of essentially all the deep puzzles still confronting them, and therefore the discipline beyond all others requiring emphasis and support.

The realities are of course much more complex than this, and each extreme requires comment. The fact that a process or product is labelled 'genetic engineering' or 'high technology', for that matter, does not endow it with magical properties that permit it instantly to overwhelm the marketplace and make its sponsors wealthy. In each of the industrial applications we have discussed, genetically engineered products will have to prove their superiority, one by one, over presently available alternatives. Thus for example, the genetically modified plants that contain insecticide-producing genes may confer benefits to the growers in terms of decreasing the number of times that insecticide spraying is done, with consequent lowering of total costs of production. However these plants will only be widely sown if the crop fetches a good price and the higher costs of seed purchase do not outweigh the savings. These in turn rely on a complex interplay of market factors. Companies producing seeds for non-transgenic plants might lower their prices to make them more attractive to farmers. The costs of processing transgenic material might be higher because of regulatory decisions concerning labelling of these products. Consumers might be unwilling to purchase products using GMO material. Any of these could affect how widely used the product will be. The possibilities also illustrate that there are considerable commercial risks for the development of any new product.

Successful commercialization will undoubtedly be a tough battle, in which there will be a few winners and many losers. Observers of Silicon Valley rarely point out that for every successful Californian electronics company there were literally scores of failures, and there is no reason to believe that it will be any different for the DNA industry in Genome Valley. While the widespread publicity about genetic engineering has achieved a highly worthwhile raising of public consciousness about commercial possibilities, the important thing for the immediate and more distant future will be to engage in meticulous and realistic case-by-case analysis of opportunities against a background of justifiable long-term optimism.

The dangers foreseen from genetic engineering will continue to prove groundless

What about the problems that could result from the use of products and therapies produced by genetic engineering? We feel sure that there will be many examples of new therapies and products that will be tested in the next decade, ranging from new drugs for heart disease, new vaccines, therapies to replace defective genes, to modified plants and animals. There will be issues of various sorts connected with each new development and there will undoubtedly be specific problems that will arise. For example, the discovery and production of erythropoietin by genetic engineering techniques have provided us with a valuable therapy to improve the quality of life of patients with kidney failure. However this product is so expensive that almost all such patients living in developing countries are denied this therapy because of its cost. Further it has become a favourite method of cheating for many athletes in endurance sports such as cycling. None of these difficulties existed before the introduction of this drug. More serious are the many suggested dangers of the production and potential release of a variety of genetically manipulated organisms, including plants, animals and bacteria. We have already covered many of the issues concerning the efficacy of existing safeguards in Chapters 9 and 10. Ecological or epidemic disasters seem most unlikely, although less serious problems could well arise. However, it is worth keeping in mind that evolution has had a very long time to create the present, constantly changing but nevertheless balanced and extraordinarily diverse biological landscape of this planet. It has done so by ensuring that DNA can change down the generations. Copying errors can be made, changing individual genes. New genes can be introduced into cells from the outside by viruses or other vectors. We now recognize that very large chunks of DNA pass from organism to organism, particularly in bacteria by a process called 'lateral transfer'. Genes can jump around within a cell. Genes are constantly being placed into novel constellations through the processes of sexual reproduction. This has sometimes given us species that startle us with their capacity to cause havoc, such as the bacteria of bubonic plague, or the grasshopper devastating whole tracts of crops. Yet even the worst examples of what evolution can do to place the ecosystem at the mercy of a single species reach their limits in space and in time. The plague did not destroy Western civilization. The grasshopper did not put an end to planned agriculture. While not doubting the capacity of genetic engineering to come up with species with quite amazing qualities, we doubt the capacity of any such species totally to wreck the world. The feedback loops within the ecosystem are too many and varied for that to happen. There is only one species that can wreck the world, and that is humans themselves, with the nuclear weapons they have invented. Human activity has caused the pollution which threatens many ecosystems; human selfishness and thoughtlessness has placed biodiversity at risk and rendered many species extinct. By comparison, the dangers from genetic engineering seem remote.

This brings us to the second set of alleged dangers. We argued in Chapter 10 that scientific results and insights were not dehumanizing, but rather energized much of what is best in humanity. Here we want to go one step further, to suggest that a fuller understanding of the biological nature of humans, not by a few experts, but by large masses of people, could prove to be a very liberating influence. Ignorance, superstition, fear of unknowable dark forces, oppression by the few gifted with knowledge and power—these have been the impediments that over the centuries have fettered the human spirit. A conviction that even the most profound and obscure realities—the nature of consciousness, the uniqueness of each individual—are the results of orderly processes, which obey rules and possess structure, must allow a person to confront his or her destiny with a heightened awareness and strength. If, further to that, a belief grows that these rules and structures are knowable, this surely permits people to walk into the future tall and free-striding, more determined to shape that future themselves. As the biological basis of the phenomenon of humans is gradually revealed, far from leading to a sterile or uniform vision, a richly patterned, fine-grained mosaic will emerge, dazzling in its complexity, diversity and subtlety. We are a long way from that point, groping about as we are at this taxonomic and descriptive stage of gene research. We cannot yet pick Einstein from a dullard or Mozart from a tone-deaf philistine on the basis of DNA sequence data. But we have made a beginning. We can describe the differences between the genetic make-up of two people more fully than ever before. Now that the first draft of the human genome has been released, the even more ambitious Human Genome Diversity Project will swing into full gear. This seeks to determine all the possible variations of given genes, and to associate these variations with observable differences between individuals. As a result, we will be able to say profound things about the diseases each is more likely to get, and the biochemical weaknesses each is capable of passing down to offspring. Of necessity, our first concern has been with abnormalities, potentially capable of detection, but analysis of variations between normal people, as a problem in its own right, has also begun. The correlation of these with those physical or mental characteristics that matter to us will be an awesome task, the work of centuries rather than decades, but we certainly need not fear what the search will uncover.

Genetic engineering has revealed the universality, beauty and order of the rules of biology. These rules create the diversity of life that we revere and cherish. It now appears a matter of prime importance that the central truths which are emerging find their way into the school curricula, not only for that small proportion of children and youths who choose to study biology as a specialty, but indeed for all future citizens. In the deepest sense, DNA's structure and function have become as much part of our cultural heritage as Shakespeare, the sweep of history, or any of the things we expect an educated person to know. The era in which the fundamentals of molecular biology and genetic engineering are taught in high school could well prove to be the era when humanity finally becomes comfortable with science, one of its prize creations, and more mature in ensuring that its power is harnessed towards noble ends.

Molecular biology—the king of the biosciences?

The third set of futurologists, coming from within science, tend not to air their speculations within public forums, but rather to exert influence within the system in an unarticulated but nevertheless effective way. We speak of the decision-makers, the trendsetters and the role models leading science from within the academic community. There is the widespread assumption that molecular biology has won pride of place as the most important discipline within science. To be a molecular biologist is best of all; if you are some other kind of biologist, you had better find yourself a molecular biologist as a collaborator, or you will be left behind in the race. Molecular biology is to be valued above all because it is the great harmonizing and welding force in biology that unifies all the specialties; and, of course, genetic engineering is the key tool of the molecular biologist.

There is much of merit in such an analysis, but also much to doubt and a little to fear. Let there be no question about the critical role which molecular biology and genetics are playing as tools in the great adventure of contemporary bioscience. Whether you are in medicine, veterinary science or agriculture; whether you seek pure biological knowledge or some practical goal; whether your specialty involves the brain or the liver or the immune system; the problem you are addressing will at some level involve cells and the intricate communication between cells. These processes involve protein molecules. Understanding them means understanding protein structure, interfering with them for good or ill means introducing drugs or other interventions which mimic or impede specific proteins. Genetic engineering, because of the superior technologies invented over the last decade, gives us the power to study DNA, RNA and protein structure in an entirely new way and with breathtaking speed. It has superseded many other ways of approaching biological problems, and immensely complemented many others. If there is any prediction about more distant horizons which can be made with complete certainty, it is that genetic engineering will yield basic information about diverse biological systems which will more than justify the resources devoted to it. Perhaps its contribution to fundamental research will prove to be its main gift.

Even though scientific research is changing, and its technological complexity is increasing phenomenally, research is principally about ideas. Advanced techniques are absolutely essential tools, and the prizes in science may well chiefly fall to those hands that master the most tools, or deploy them most effectively; but the logical and imaginative constructs that human minds produce when the tools have done their work are the essence of science, and that which distinguishes it from technology (or research from development). It is therefore important to remember that genetic engineering is a tool. In biology, as in physical sciences, the complex real-life situations which we seek to comprehend can be viewed from many perspectives. Let us take a simple example from medicine, say a patient with multiple sclerosis. The neurologist will look at this as a problem in diagnosis and therapy; the perspective will be analysis of symptoms, signs and diagnostic tests. Given the uniqueness of that particular patient's case, one or other remedy of an available set will be chosen. The clinical immunologist, on the other hand, is more concerned with causative mechanisms: is the disease due to a virus gone underground, or to an inappropriate immune attack or both? He or she, then, turns to any one of a number of neuroscientist colleagues. One will investigate in detail the impaired nerve conduction, using complex electronic recording apparatus. Another will study antibodies present in the patient's serum, to see if some capable of damaging nerves are present. A third will look to see if the patient has certain genes known to increase the risk; a fourth will be more concerned to determine in detail what molecules on the nerve have been attacked by the ravages of disease. But beyond the neurologist and the neuroscientist, 'above' and 'below' them, stand others involved in the study of the multiple sclerosis problem. 'Above' we find the epidemiologist trying to make sense of the peculiar geographic distribution of the disease, and the expert in community medicine working out how best to look after the patients, given the limited resources available. 'Below' we have the full spectrum of fundamental biosciences, as these may impinge on brain structure, organization and function.

The point of this simple example is that each of the relevant perspectives is a valid and valuable one. Within each frame of reference, new knowledge will accrue, new insights will enhance capacity to intervene and finally, one day, the puzzle will be solved and the human energy thus released will attack other problems. Each of the levels of search and striving, while seeking all manner of ways to communicate with the level immediately 'up' or 'down' from it, will preserve its own idiom and its own disciplinary integrity; and each one will reserve a special place for the new problems posed as each solution comes to hand. For this multi-dimensional network which is world science, we see nothing but limitless growth into the indefinite future. If this is so, it is also important to preserve balance and order between the disciplinary layers. It would make no more sense to plough all new resources in the multiple sclerosis field into epidemiology than it would to channel them into immunology or electro-physiological diagnostic equipment.

The trick, therefore, for world biological science will be to exploit the truly wondrous potential of genetic engineering without doing harm to other fertile disciplines. Many of the exciting problems on which genetic engineers work stem from within their own framework, but many more come from the outside. Cancer biologists, immunologists, endocrinologists, neurobiologists, developmental biologists and many others are crying out for collegial assistance from genetic engineers, so much so that there is a slight danger of transmitting to the younger generation the signal that this is the only way to go. Yet most of the phenomena for which we seek molecular explanations will continue to come from outside sources, and we need scientists committed to (and steeped in the folklore of) the 'higher' disciplines to provide a continuing stream of cannon-fodder for the genetic engineers.

Why do we term an exaggerated commitment to genetic engineering only a slight danger? Because, all in all, we believe the world science system to be in fine condition, full of all manner of self-correcting mechanisms. That is not to say it could not be improved. We hear much about the fierce competition for grant funds, particularly among younger research workers not yet established in a career structure. Scientists are apprehensive about new intrusions by industry, dissatisfied by treatment in the media, worried about over-regulation, always scornful of politicians' lack of understanding. Yet, beyond all that, there is an excitement and an elation that is barely containable. Biology is in the midst of an explosive leap, one which will make decades of work for future historians. It is a great time to be young in biomedical science.

What will become of humankind?

Before we conclude, it perhaps worth asking what the further future holds in store for humanity. Of course this must be speculative in the highest degree, but given the technology we have in hand, what will happen? In the near term, it is almost certain that the next ten to thirty years will see the production of many new drugs aimed at many complex diseases such as heart disease, rheumatic disorders and diabetes, as well as the amelioration and reversal of aging. As the 'baby-boomers' age, the market forces that induced drug companies to develop drugs such as Viagra, will lead to concentration of research efforts on ways to turn back the biological clock and extend the span of human life. Gene therapy techniques will decrease the toll taken by genetic diseases and we will probably stand on the brink of developing replacement organs and tissues. Genetic profiling of people for diseases of which they are at risk will be a routine part of medical practice. These efforts of course will most affect the favoured few who have the means to purchase the fruit of this research. However the lot of humankind in the developing countries will improve as the successful development of vaccines against diarrhoeal and respiratory diseases and malaria slashes death rates, particularly among the young. Although populations will increase, the development of new plants in particular with superior nutritional characteristics will help to reduce malnutrition. The rate of land degradation will lessen as plants are bred that are tailored to live in these harsher environments with less water, fertilizer and insecticide usage. Transgenic animals will become factories for the production of many new proteins of medical importance. Cloning technology will rapidly produce flocks of these animals. This future is one we can glimpse dimly now, but what may take place even further on?

Perhaps the most surprising developments will take place from the fusion of genetic engineering with the other great revolution of the twentieth century—computers and information technology. The rapid pace of progress in computing rivals that of biotechnology and we can already see in the not too distant future the prospect of semi-intelligent, or perhaps even truly conscious machines. Using fabrication technology developed by the computer industry, engineers are using photolithographic techniques to etch microscopic pieces of silicon into rudimentary pieces of machinery, such as gears, levers and motors. It is thought that tiny machines, perhaps able to enter the body and perform surgery, will eventually be the result of this area of endeavour called nanotechnology. Robots are becoming increasingly sophisticated and Aibo ERS-111, the first robot pet, was released on the market in 1999 by the Sony Corporation.

The first meetings of silicon and carbon are being recorded in the scientific literature. We have spoken about the discipline of bioinformatics and the absolute reliance placed on computers to store and analyse genomic sequences. In 2000, scientists reported using DNA strands and the way they are synthesized in solution to form a sort of organic computer that can solve simple problems in mathematics, in this case the possible journeys taken by a knight to cover the chessboard. The ultimate aim is to develop biomolecular computation (BMC) employing biotechnology to do massive parallel computing at the molecular scale (also known as 'DNA computing'). Protein molecules themselves are being incorporated onto chips to determine if their changes in shape can be used to store information. The goal here is that you might be able to construct a 'biological hard disc' with incredible storage capacity, say a device the size of a sugar cube that stores a terabyte of information. This mix of engineered species, tiny machines, robots and artificial intelligence is a heady brew that fascinates scientists, science fiction writers and the public alike. When we come to revise this book in the coming years, it is perhaps here that we will find the greatest surprises, where life as we currently define it is truly reshaped.

Appendix
Subcellular organelles

Rather than being an amorphous bag of proteins, fats, and carbo-hydrates, the living cell is a highly organized structure with discrete organelles undertaking specialized functions. Some of the more important organelles are described below.

The mitochondria

The mitochondria are the powerhouses of the cell, creating little mobile packets of energy for use in the myriad chemical conversions that each cell must perform each minute. The process used is called oxidative phosphorylation, but this daunting term need not deter us because what we need to know is quite simple. When you burn a log of wood, you are performing a relatively uncontrolled act of oxidation. The carbon and hydrogen in the wood combine with oxygen to yield carbon dioxide gas, water and energy by way of heat. The energy is rapidly dissipated, but the mitochondria have figured out a way of promoting oxidation so that most of the energy generated is chemically stored. The controlled oxidation of cellular foodstuffs leads to the build-up of energy-rich, phosphate-containing molecules called adenosine triphosphate or simply ATP. Whenever the cell needs a bit of energy, it grabs a few of these stored ATP molecules and breaks them down to adenosine di-phosphate or ADP. The energy stored in the energy-rich phosphate bond is released to drive the required chemical conversion, where-upon the hard-working mitochondria reconvert the ADP to ATP, building back the energy store.

The lysosomes

The lysosomes are central to the digestive system of a cell. Cells take up a wide variety of substances from their environment. Some cells, such as the scavenger white cells of the blood, are big eaters and swallow particles such as invading microbes or bits of dead cells. Others take up only nutrient molecules present in the fluid around them. In either case, it is necessary to break down the material entering the cell into portions that can be used in cellular metabolism. This requires the action of digestive enzymes, and, as in the human stomach, an acid environment. If such enzymes and acids are spilt freely into the cytoplasm, the effects could be disastrous. Therefore, the cell keeps the enzymes and acids tucked away behind a barrier inside the lysosomes. The nutrients are at first also kept in little bags or pouches called vacuoles. The lysosomes discharge their contents into these food vacuoles so that digestion can proceed.

The ribosomes

Scattered throughout all cells there are tiny punctate organelles called ribosomes. Frequently these appear in aggregates of ten or twenty and then they are referred to as polyribosomes. The ribosomes are central to our story, because they are instruments involved in reading the genetic code, making the proteins which end up doing most of the work of the cell.

The endoplasmic reticulum

When the cell makes proteins for its own internal use, the ribosomes concerned lie free inside the cytoplasm. However, many specialized cells make proteins which have to leave the cell and circulate in the blood to be of use to the body. A hormone like insulin is a typical example. Insulin is made by cells in the pancreas and regulates metabolic processes in virtually every cell of the body. Antibodies illustrate the point equally well. Antibodies are the proteins which recognize and neutralize foreign invaders that enter the body. They are made by a specialized cell type called a plasma cell, and are secreted by it, but may act anywhere that the bloodstream takes them. The endoplasmic reticulum is a kind of river system that guides the molecules destined for export to an appropriate packaging centre. The endoplasmic reticulum is made up of double membranes between which the protein molecules flow. The polyribosomes making proteins destined for export are actually attached to these membranes so that there can be no doubt about the newly synthesized export product getting to the right place.

The Golgi apparatus

Proteins destined for secretion move to a system of membranes and sacs or pouches. This system sits in the cytoplasm right next to the nucleus. It is called the Golgi apparatus and is a concentrating and packaging centre. When the proteins are ready for export from the cell, they move in tiny bubbles from the Golgi to the surface of the cell, where they are released. At various points in the journey from endoplasmic reticulum to outside the cell, sugars may be linked on to the proteins, an important point for our later consideration, as genetically engineered bacteria fail to perform this function.

The cell membrane

Though not an organelle, the outer skin of the cell, the cell membrane, requires special comment. Chemical reactions inside the cell occur in a watery environment, and there is also a watery milieu outside the cell called the extracellular fluid. What separates the molecules inside the cell from the outside world is a skin largely made up of fats or lipids. This skin, the cell membrane, is impermeable to most molecules, thereby preserving the cell's autonomy and integrity. Obviously, therefore, when molecules need to get in, to feed the cell or deliver messages, or when molecules need to get out, say to provide the bloodstream with some secretion, special channelling devices of various sorts need to be created. The membrane is therefore a very dynamic and functional entity. Floating in the sea of fat we find a large number of proteins, and some of these function as sensitive antennae, constructed to receive signals and messages of various sorts from neighbouring cells or from hormones in the bathing fluid. These molecules are called cell membrane receptors.

Glossary

adenine   One of the small molecular building blocks, called bases, which make up the coding units of DNA and RNA. Often abbreviated to A. In DNA, pairs with thymine (T).

adenosine diphosphate (ADP)   A molecule consisting of adenine plus a sugar plus two phosphate groups important in the energy economy of a cell. Oxidation of fuel molecules such as glucose permits ADP to take up an extra phosphate and thereby to trap energy.

adenosine triphosphate (ATP)   A molecule consisting of adenine plus a sugar plus three phosphate groups which acts as the universal currency of free energy in biological systems. Conversion of ATP to ADP releases energy which drives the work of the cell.

adjuvant   A substance which increases the efficacy of a vaccine, i.e. stimulates the immune response.

AIDS   A disease, the acquired immune deficiency syndrome, in which certain T lymphocytes are destroyed and the patient is left defenceless against infections.

allosteric effect   A protein molecule is caused to change shape through union with another molecule. As a result, a new active site is exposed.

alpha-foetoprotein test (AFP)   A test administered during pregnancy to measure the amount of a foetal protein in the mother's blood. Abnormal amounts of the protein may indicate genetic problems in the foetus, such as an abnormal spinal cord.

amino acid   Building block of proteins. There are twenty naturally occurring amino acids.

amniocentesis   A prenatal test in which cells surrounding a foetus are removed in order to examine the chromosomes.

antibody   A special protein molecule made by the immune system of vertebrate animals, specifically tailored to fit other molecules, for example bacterial toxins, much as a given key fits a particular lock.

antigen   A generic term for a molecule with which an antibody reacts.

artificial insemination   The injection of semen into a female's uterus (not through sexual intercourse) in order to make her pregnant.

autoimmune disease   A disease in which the body manufactures antibodies against some component of itself; thus a form of civil warfare in the body, one cell attacking another.

bacteria   Very small, single-celled life-forms that reproduce quickly.

bases   Distinct chemical ingredients that are the building blocks of the genetic material of all life-forms—DNA and RNA.

biotechnology   The use of living things or components from living things to make a variety of products.

cDNA   A stretch of DNA synthesized by enzymes as a faithful copy of a particular stretch of RNA which thus preserves the information content of that RNA.

cell   A fundamental organizational unit of all living matter. The simplest forms of life consist of just one cell, e.g. bacteria, algae or certain parasites. Higher life forms are multicellular organisms, permitting specialization of cellular function, i.e. a division of labour between cells.

cell membrane   The fatty outer skin of a cell which separates it from the next cell, from the fluid bathing cells, or from the environment.

cell membrane receptors   Protein molecules, frequently with some sugars attached, which reside in the cell membrane and possess the capacity to bind specifically some molecule which floats past, e.g. a hormone, a nutrient, or a trigger for cellular activation.

chain   Used in the context of a chain of amino acids which follow a sequence determined by the gene for that chain; many proteins consist of two or more chains linked together chemically. Thus insulin has an α and a β chain; many anti-body molecules have four chains, two smaller ones called light, and two larger ones called heavy. Usually accompanied by the adjectival noun polypeptide (q.v.) meaning many amino acids.

chorionic villus sampling (CVS)   A prenatal test in which cells surrounding an embryo are removed in order to examine the chromosomes.

chromosome   A very long double-stranded DNA molecule packed together with certain proteins, forming a sausage-like entity readily seen under the microscope when a cell divides. The number of chromosomes per cell is a characteristic of a species; thus humans have forty-six chromosomes per cell.

cloning   Causing asexual division. Frequently used as jargon in genetic engineering to describe the sequence of events by which a gene is caused to replicate a large number of times in some foreign host cell. Also often used to mean to make an exact copy of.

codon   A sequence of three bases of DNA or RNA which codes for one amino acid.

colony   A clustered group of cells which arose from a single cell by asexual division, thus a bacterial colony may be a visible spot 1–2 millimetres in diameter consisting of millions of bacteria that are growing in a jellified medium.

colony stimulating factors   Proteins that stimulate the development of new white blood cells. Named for the characteristic clump of new cells seen when white cells are grown on agar plates in the presence of the factor.

cosmid   A virus-like vector used by genetic engineers that combines some of the advantages of phages and of plasmids as instruments for the cloning of genes.

cystic fibrosis (CF)   A genetic disorder affecting the mucus lining of the lungs, leading to breathing problems and other difficulties.

cytoplasm   That portion of a cell which is not the nucleus; the site where proteins are made and where chemical energy is generated; the 'factory' portion of the cell.

cytosine   One of the four small molecular building blocks, called bases, which make up the coding units of DNA. Often abbreviated to C. In DNA, C pairs with guanine (G).

data bank A collection of information organized so that specific facts can be retrieved as needed. All gene sequences are stored in data banks such as Genbank or EMBL.

differentiation The process whereby cells gain more specialized function. Thus, as a cell destined to turn into a red blood cell gradually builds up more and more haemoglobin, it is said to differentiate.

diploid Refers to the number of chromosomes in a cell or a set. Normal cells contain chromosomes in pairs. Thus twenty-three pairs make up the forty-six chromosomes in a normal diploid human cell. Cancer cells are frequently hyper-diploid, i.e. contain more than forty-six chromosomes.

disulphide bond A chemical linkage between two sulphur-containing amino acids either within a single polypeptide chain or between the component chains of a multichain protein. The disulphide bonds stabilize the shape of a protein and help to keep multichain proteins as a single molecule.

DNA Deoxyribonucleic acid. A double helical molecule consisting of a sugar-phosphate backbone and a sequence of base pairs constituting the coding units of the genetic code. Particular stretches of DNA constitute a gene, one gene being that stretch which encodes one polypeptide chain.

DNA fingerprinting The analysis of sections of DNA for purposes of identification.

DNA ligases Enzymes which catalyse the formation of the chemical bonds needed to weld pieces of DNA together. Thus, DNA ligases may join a gene from an animal cell with DNA from a phage virus, creating recombinant DNA.

DNA marker A gene or other fragment of DNA whose location in the genome is known.

E. coli *Escherichia coli.* A harmless bacterial species which resides in the human intestine. Frequently used in genetic research, e.g. as a host cell for phages or plasmids carrying recombinant DNA.

electrophoresis A procedure in which a mixture of molecules is subjected to an electric current ensuring that each molecule moves at a rate influenced by its net electric charge; thus a useful way of analysing and separating complex mixtures of molecules, e.g. proteins.

embryo The early stage of development of an animal before birth. In humans, the embryo stage is classified as the first three months following conception.

endonuclease An enzyme capable of cutting DNA.

endoplasmic reticulum A system of channels inside the cytoplasm of a cell for the assembly and export of protein molecules.

endotoxin A molecule derived from the cell wall of bacteria which is highly toxic to animals.

enzyme A protein molecule capable of catalysing chemical reactions within the body. Enzymes are strategic components of all cellular metabolism.

erythropoietin (EPO) A naturally occurring protein in the body that stimulates the production of red blood cells.

evolution The process by which all forms of plant and animal life change slowly over time because of slight variations in the genes that one generation passes down to the next.

exon That portion of the gene which encodes a portion of the amino acid sequence of the protein. One gene may contain several exons.

expression vectors Tools of the genetic engineering which permit a gene to be inserted into a cell in such a manner that, on appropriate signalling, the cell will manufacture large amounts of the protein for which that gene codes.

foetus An animal in the later stage of development before birth. In humans, the foetal stage is from the end of the third month until birth.

gel electrophoresis A procedure in which a mixture of proteins, nucleic acids or other molecules is made to penetrate into a jellified medium under the influence of a strong electric current. Molecules migrate at a rate dependent on their net electric charge and, on this basis, different molecules can be separated from one other.

gene A stretch of DNA containing specific information for the construction of one polypeptide chain or protein. In higher organisms, genes consist of exons and introns, q.v.

gene activation A process in which a command is given which ensures that messenger RNA molecules will be made as copies of the particular gene being activated. Thus, gene activation is the first step in protein synthesis.

gene therapy   The altering of genes in order to affect their function.

genetic code   The code whereby the structural information for proteins is encoded in the nucleotides of the DNA. Proteins are strings of amino acids, one amino acid out of twenty being chosen for each spot in the string. Nucleic acids are strings of nucleotides, one nucleotide out of a possible four at each spot. A sequence of three nucleotides specifies one amino acid.

genetic engineering   The technology by which genes can be isolated, transferred to other cells, replicated and activated.

genetic fingerprinting   A method that provides a unique identifying pattern of bands on a Southern blot for each individual. The technique is based on the fact that everyone's DNA (except for identical twins) has some different bases or arrangements of sequences.

genetic linkage study   Examination of the DNA of family members to determine who may be at risk for a genetic disorder occurring in the family. Doctors look for variations that consistently appear in the DNA of family members with the disorder. These DNA variations may or may not be related to the genetic disorder. However, if they appear in the DNA of another family member, it can indicate the person's risk of inheriting the disorder.

genetic profile   A collection of information about a person's genes.

genetics   The field of science that looks at how traits are passed down from one generation to another, through the genes.

genome   A noun used to denote the total complement of genes in a cell or individual.

germ cells   The cells of the body involved in reproduction. Sperm of the male and eggs of the female are formed from germ cells.

germ-line therapy   The altering of genes in reproductive cells (sperm or egg) which will result in these changed genes being passed on to any offspring that may be created.

Golgi apparatus   A packing centre for the concentration and temporary storage of protein molecules destined for export by the cell.

growth factors   The collective name for a group of proteins that stimulate the division of cells. Each growth factor stimulates production of particular cell types.

guanine   One of the small molecular building blocks, called bases, which make up the coding units of DNA and RNA. Often abbreviated to G. In DNA, pairs with cytosine (C).

haemoglobin   An iron-containing pigmented protein contained in the red blood cell which is responsible for carrying oxygen around the body and releasing it for the use of the cells.

haemoglobinopathies   A group of diseases resulting from an abnormality in the gene for one of the chains of haemoglobin.

haemophilia   A heritable disease in which a protein essential for blood clotting is defective. Patients bleed too readily, particularly after injury.

haploid   Refers to the number of chromosomes in a cell or set. Most cells contain pairs of chromosomes, known as a diploid set, but the cells for reproduction, the sperms and the ova, contain only half this number, e.g. twenty-three chromosomes in the human, instead of forty-six in other cells. This constitutes a haploid set. The number is restored to forty-six when sperm and egg fuse.

heredity   The handing down of certain traits from parents to their offspring. The process of heredity occurs through the genes.

homopolymer tailing   A procedure by which a string of nucleotides, all the same, is added to the end of one strand of a DNA molecule. This string, e.g. A–A–A–A–A–, will readily stick to another DNA molecule tailed with the complementary nucleotides, e.g. T–T–T–T–T.

hormone   A chemical messenger molecule, travelling in the bloodstream, synthesized by cells in an endocrine gland and capable of influencing growth and metabolism within other, perhaps distant, cells which possess receptors for that hormone.

human genome project   An ambitious plan by scientists that has obtained the base sequence of all the DNA in humans—some three billion bases.

hydrogen bonding   In a hydrogen bond, a hydrogen atom is shared by two other atoms (in biological systems, nitrogen and oxygen). Hydrogen bonds stabilize the structure of proteins and of DNA.

insulin   A hormone made by β cells in the pancreas necessary for the proper utilization of glucose within the body.

interferon   A generic term used to describe three groups of molecules, the α, β and γ interferons. These molecules are synthesized by cells as a result of virus infection and temporarily interfere with the growth of other viruses in that or nearby cells.

intron   Stretch of DNA occurring within a gene which, however, does not code for amino acids of the relevant protein. When a gene is activated, the RNA molecules made as copies of the gene faithfully reflect both introns and exons (q.v.) but before this RNA travels to the cytoplasm, the sequences corresponding to introns are cut out and the (shorter) RNA corresponding only to copies of exons is joined up to constitute the final messenger RNA template.

***in vitro* fertilization**   The mixing of eggs with sperm in a laboratory dish in order to achieve conception.

karyotype   A picture of the chromosomes in a cell that is used to check for abnormalities. A karyotype is created by staining the chromosomes with dye and photographing them through a microscope.

lac operon   A group of genes and control elements responsible for the proper utilization of lactose by bacterial cells. Frequently used by genetic engineers as a switching device for gene activation, q.v.

lipase   An enzyme capable of catalysing the digestion of fats.

lipid   A technical term for describing fatty molecules in biology.

lysosomes   Small pouches within the cytoplasm of cells containing enzymes capable of digesting particles or molecules that enter the cell.

major histocompatibility complex   A group of genes which determine the tissue type of an individual, i.e. compatibility with another for organ transplantation. Also involved in the regulation of immune responses.

meiosis   A special type of cell division which creates the reproductive cells, the sperm and the ova. During the process, not only is the number of chromosomes halved, e.g. in the human from forty-six to twenty-three, but also the paternal and maternal genes become recombined in new ways. As this happens differently in each meiotic division, no two sperms or no two ova in any individual are exactly the same.

messenger RNA   A copy of the DNA which moves from nucleus to cytoplasm and serves as the immediate coding entity which is decoded as proteins are made.

mitochondria   Subcellular particles within the cytoplasm which generate chemical energy for use by the cell in a large number of bioconversions.

mitosis   The non-sexual division of cells whereby each daughter cell receives the full diploid number of chromosomes.

mobile genetic element   Portion of the genome which, unlike most DNA, does not occupy a fixed position but can jump from spot to spot on a chromosome or even move between chromosomes.

molecule   A grouping of atoms which together make a stable substance.

monoclonal antibody   An antibody made by the progeny of a single cell, thus extremely pure, precise and homogeneous.

mutation   Changes that occur to bases appearing in the DNA inside a cell. The changes may be to substitute one base for another, or to add or remove a base

Niemann-Pick disease   A rare genetic disorder in which a defect in a gene inside the lysosome causes accumulation of lipid inside the cell. Somewhat related to Tay-Sachs disease, q.v.

nucleic acids   Two types of polymer molecules, DNA and RNA (q.v.) which act as the repositories of genetic information. They consist of a backbone of alternating sugar and phosphate portions, with a coding unit attached to each sugar.

nucleotides   The building blocks from which the polymeric nucleic acids are made, i.e. a sugar with an attached coding unit and phosphate group.

nucleus   The control centre of the cell, where the DNA resides, separated from the 'factory' portion of the cell, the cytoplasm, by a double membrane.

oncogene   A gene or genes which, when inappropriately activated, can be involved in the production of cancer.

organelles Small subcellular particulate structures within the cytoplasm of a cell, recognizable in the electron microscope and frequently separable from other organelles or the fluid, structureless part of the cell by biophysical techniques. Many organelles possess specific functions known in detail.

palindromic sequences Stretches of DNA which read identically backwards or forwards.

peptide synthesis The process by which amino acids are joined together to form short or long chains.

phage Abbreviation of bacteriophage virus, a virus capable of infecting and destroying bacteria. Frequently used as a vector (q.v.) by genetic engineers.

phosphorylation The metabolic process whereby a phosphate group is added to a molecule.

plaque A clear area, e.g. where a phage population has destroyed bacteria growing on a jellified medium.

plasmid A circular piece of DNA capable of self-replication within a cell independently of nuclear DNA. Frequently used as a vector (q.v.) in genetic engineering.

pluripotent Capable of giving rise to most tissues of an organism.

polymer A molecule made up of a number of smaller subunits.

polymerase chain reaction A method for rapid amplification of a specific stretch of DNA in the test tube. This powerful technique has many applications in research and medicine.

polypeptide A stretch of amino acids constituting a protein or one chain (q.v.) of a multichain protein.

polyribosome A collection of ribosomes (q.v.) attached to a messenger RNA molecule engaged in aiding the synthesis of proteins according to the coded instructions in the RNA.

primary transcript That molecule of RNA first synthesized as a faithful copy of a whole gene when a gene is activated. Portions of the primary transcript are cut out before the messenger RNA moves to the cytoplasm.

probe A stretch of DNA or RNA labelled with a radioactive isotope, capable of binding to, and thus 'finding' a stretch of DNA with a complementary sequence.

protein Molecules composed of amino acids and which perform most of the cell's work. Includes enzymes, hormones, antibodies, carriers for other molecules, receptors and structural molecules.

protein kinase An enzyme which catalyses the addition of a phosphate group to certain amino acids of proteins.

protein synthesis The process by which the amino acids are joined together to form proteins. Almost synonymous with peptide synthesis, except that the latter usually refers to shorter stretches of amino acids.

proteome A term used to denote the total complement of proteins found in a cell or individual. It includes all the different variants of a single protein that may result from modifications such as the addition of phosphate. The study of the proteome is called proteomics.

reproductive technology The application of scientific knowledge to assist in making babies.

restriction endonucleases Enzymes which cut the DNA double helix only when a particular sequence of base pairs is present.

retrovirus A virus which uses RNA as the genetic material but possesses the enzyme reverse transcriptase (q.v.) and which can thus cause a DNA copy of itself or some part of itself to be made inside the cell.

reverse transcriptase An enzyme capable of using RNA as a template and creating a DNA copy of the relevant sequence.

ribosomes Small, particulate entities within the cytoplasm which attach to messenger RNA and help to translate that message into a particular amino acid sequence. Essential for protein synthesis in the cell.

RNA Ribonucleic acid. A single-stranded molecule consisting of sugar, phosphate and a string of bases. Different sorts of RNA have different functions. Messenger RNA is the immediate template for protein synthesis.

selective breeding The selection of certain seeds or animals for reproduction in order to influence the traits inherited by the next generation.

somatic cell A cell of the body other than egg or sperm.

somatic cell nuclear transfer The transfer of a cell nucleus from a somatic into an egg from which the nucleus has been removed.

species A single, distinct class of living creature with features that distinguish it from others. Sometimes defined as one of a group of organisms which can interbreed.

stem cells   Cells that have the ability to divide without limit and to give rise to specialized cells.

sticky ends   Short single-stranded sequences of DNA capable of binding to short, complementary stretches on other DNA molecules.

substrate   The target for an enzyme's action.

Tay-Sachs disease   An inherited disease, occurring predominantly in Ashkenazi Jews, due to a genetic defect in an enzyme, hexosaminidase A, which leads to abnormal accumulation of certain fats in nerve cells causing severe mental retardation and death.

thymine   One of the small molecular building blocks called bases, which make up the coding units of DNA and RNA. Often abbreviated to T. In DNA, pairs with adenine (A).

tissue typing   The process by which scientists determine the genes of a person which are important for organ transplantation.

totipotent   Having unlimited capability. Totipotent cells have the capacity to develop into extra-embryonic membranes and tissues, the embryo, and all tissues and organs.

transcription   The process whereby the DNA double helix unwinds and an RNA copy of a gene is synthesized complementary to one of the strands.

transfection   Insertion of DNA into a cell without a vector and integration of that DNA with the cell's own genes. Generally an inefficient process but occurs sufficiently frequently that, if transfected cells can be selectively grown, genetic engineering can be achieved.

transfer RNA   An abbreviation of amino acid transfer RNA. Each particular transfer RNA molecule can ferry a particular amino acid to the right spot on the ribosome, thus helping in protein synthesis.

transgenic   Containing genes from another species.

translation   The process by which the coded message in messenger RNA is read, resulting in the formation of a corresponding protein.

transposons   Mobile stretches of DNA which can move around within the genome instead of (like most DNA) residing in one place in the one chromosome.

uracil   A base unique to RNA, informationally equivalent to cytosine in DNA. Abbreviated to U.

vaccine   A preparation used in the immunization of a person or animal, designed to protect the immunized individual from a particular virulent infection.

vacuole   A sack-like subcellular entity in a cell which looks relatively translucent in the electron microscope. Frequently involved in transporting food into the cell or some secreted product out of the cell.

vector   A tool of the genetic engineer used to transport recombinant DNA into a host cell and to permit its extensive replication there, independently of the replication of the cell's own DNA; a generic term covering phages, plasmids, cosmids and other types of mobile DNA.

virus   The smallest and simplest forms of life: micro-organisms which are made merely of a protein shell and a genome. A virus reproduces by inserting its genome into the cells of other life-forms. As those cells duplicate, so does the virus.

Suggestions for further reading

We live in an age where there are enormous amounts of information available on virtually any topic. Much of this can be accessed by the World Wide Web, and there are a number of sites that we think are of value. The Internet is an intensely democratic institution and anyone, from those in large corporations or scientific institutes to private individuals, can put their site up and present information. Beware, however, that the quality of information can vary widely from the true to the fantastic. For those who prefer their information in printed form, we also include a brief list of books.

To help the reader we have created the Reshaping Life site where we will keep an updated set of links to these and to other sources of information on genetic engineering and related issues. In addition there will be new material on evolving issues of interest.

Reshaping Life
<http://www.med.monash.edu.au/reshapinglife>
    Username: life
    Password: reader
The username and password should be entered in lower case letters.

As for books, we provide just a brief listing of some classic titles here; there is a more extensive list on the Reshaping Life web site. The following books have been listed in ascending order of difficulty. It is recommended that you begin at the top and work down!

Judson, H. F. (1989) *The Eighth Day of Creation* Simon & Schuster, New York, 686 pp. This book is highly recommended as a first class piece of scientific journalism. The author spent seven years researching the story of DNA, and has interviewed most of the key figures who have created the molecular biology revolution. The book is a mixture of popular science, recent history and material of considerable human interest. Parts of it are far from easy, but a little skipping here and there would not destroy the narrative flow. The book ends at the time when the genetic engineering explosion is about to begin.

Watson, J. D. and Tooze, J. (1981) *The DNA Story* W. H. Freeman & Co, San Francisco, 605 pp. This fascinating work describes the unfolding story of genetic engineering by reproducing a large number of primary documents, including press articles, private correspondence, meeting reports, congressional testimony and draft legislation. The authors have annotated the material with short essays that introduce each section, and have provided a brief, beautifully illustrated description of recombinant DNA technology.

Plant, D. W., Reimers, N. and Zinder, N. D., eds (1982) *Patenting of Life Forms* (Banbury Report 10), Cold Spring Harbour Laboratory, 337 pp. This book summarizes a conference which brought together leading lawyers and scientists interested not only in the specific subject of the title, but in broader issues of technology and the law. It forms a good starting point for readers with a special interest in the legal aspects of genetic engineering. Although now somewhat out of date, it still has some relevant sections.

Alberts, B., Bray, D., Lewis, J., Raff, M., Roberts, K. and Watson, J. D. (1994) *Molecular Biology of the Cell* (3rd edn) Garland Press, 1294 pp. This comprehensive textbook covers the working of the cell and all the molecular processes in great detail. This is a university-level text and recommended only to those who wish to track down some aspect of the field in great detail.

In a field that is moving so rapidly, interested readers who wish to keep up with the broad trends would do well to follow *Scientific American* and *New Scientist*, both highly respected publications, which frequently feature key aspects of molecular biology among their articles. *Science* magazine has an excellent section for the layperson which covers many scientific areas. Many newspapers such as the *Washington Post* have science sections accessible on the Internet.

Index

Words marked with an asterisk are included in the Glossary (pages 232 to 244).